THE

BIOLOGICAL

CLOCK

TWO VIEWS

THE BIOLOGICAL

FRANK A. BROWN, JR.

Morrison Professor of Biology
Northwestern University
Evanston, Illinois

J. WOODLAND HASTINGS

Professor of Biology
Department of Biological Sciences
Harvard University
Cambridge, Massachusetts

JOHN D. PALMER

Chairman, Department of Biology
New York University
New York, New York

CLOCK *two views*

ACADEMIC PRESS
New York
London

The design of the text and cover was created by
Wladislaw Finne.

The drawings are by Visa-Direction, Inc.

Cover photograph by Bruce Roberts from
Rapho Guillomette.

ACADEMIC PRESS, INC.
111 Fifth Avenue, New York, New York 10003

United Kingdom Edition published by
ACADEMIC PRESS, INC. (LONDON) LTD.
Berkeley Square House, London WIX 6BA

Library of Congress Catalog Card Number: 70-107551

Printed in the United States of America

CONTENTS

PREFACE

Only relatively recently have biologists
become aware of the fact that organismic
processes do not necessarily proceed at a
constant rate when studied in standard,
unvarying conditions in the laboratory.
Instead, many physiological processes have
been found to undergo cyclic changes, the
periods of which approximate twenty-four
hours in length. Because these daily changes
persist in the absence of natural day-night
alterations in the environment, organisms are
thought to possess their own physiological
mechanism for keeping time. This entity
is called the "biological clock." So far,
the exact nature of the clockworks has not
been discovered.

As a result of studying the rhythmic processes in a great variety of plants and animals, a number of common properties have
been elucidated. These properties have then
been used to construct two major, contrasting hypotheses—the exogenous and endogenous versions—of the mechanism of the
biological clock. In this volume, these
hypotheses, plus the evidence on which they
are founded, are discussed by two of the
leading proponents in the field of biological
rhythms: Drs. Frank A. Brown, Jr., and J.
Woodland Hastings.

The book is an outgrowth of the 1969
James Arthur Lecture Series on "Time and
Its Mysteries" held at New York University.
In the traditional vein, a manuscript was
produced at the close of the series to document the event. In its original form it was to
be an archival record. However, recognizing
the burgeoning interest in biological rhythms,
it was felt that with certain revisions and
additions the manuscript could become an
enlightening, succinct book, describing the
current thinking on this byway of biology.

As an organizer of the lecture series I
accepted the role of editor and also con-

tributed a short introductory section to the book in an attempt to set the stage for the Brown and Hastings essays. We hope that this volume will serve as an introduction to those uninitiated in the field and as a clarification of the major points of view on the biological clock mechanism to those already familiar with biological rhythms.

We would like to thank Mrs. Pat Dowse for the initial preparation of figures 1-1, 2-6, 2-23, 2-24, 2-26, 2-27, 2-28, 3-7, 3-10, and 3-22.

JOHN D. PALMER

INTRODUCTION TO
BIOLOGICAL RHYTHMS
AND CLOCKS

JOHN D. PALMER
Chairman, Department of Biology
New York University

Many biological processes, both at the multicellular and cellular levels of organization, undergo *regularly* recurring quantitative and qualitative changes. Highs in these processes are repeated with such *beat*like regularity (the period is 24 hours) that the processes are referred to as being *rhythmic*. Rhythms have now been described for literally thousands of organisms.

Just before the birth of Aristotle, the first written account appeared on what today we call biological rhythms; an early amateur naturalist had observed that certain plants (legumes) stand with their leaves folded to the sides of their stems at night and raise them—as if in a pagan gesture—to the sun in the morning (Figure 1-1). Day after day, throughout their entire lives, they repeat this monotonous pattern.

FIGURE 1-1

Bean seedlings with the leaves in the raised, daytime; and lowered, nighttime positions—the extremes of the sleep-movement rhythm. This unusual plant property—active movement—is brought about by a tiny package of specialized cells located eccentrically at the base of each leaf. These cells periodically inflate with water and lift the leaf in a way analogous in principle to the hydraulic piston that lifts the scoop on a bulldozer.

About 2400 years after this observation an inquisitive scientist studied these leaf movements in the laboratory and found to his surprise that even when he deprived the plants of all obvious information about the time of day (he maintained them in constant darkness and at a relatively constant temperature) that the up-and-down "sleep movements" of the leaves continued in near synchrony with their feral companions outside in the garden (Figure 1-2). This discovery clearly demonstrated that these organisms had some mysterious means of keeping time; they were described as possessing *biological clocks*. With this conclusion, the field of biological chronometry was born. During its early development it attracted the attention of such eminent scholars as Charles Darwin, Henri Dutrochet, Wilhelm Pfeffer, and Svante Arrhenius; all were men who solved major problems in their day, but who could not resolve the mechanism controlling the sleep-movement rhythms of plants.

It was only about twenty-five years ago that a real interest in biological rhythms burgeoned, and in the interval since then rhythmic physiological processes in numerous organisms—representing all the major groups of plant and animals—have been described; almost all of these rhythms were found to persist in constant conditions (i.e., light and temperature cycles were precluded) in the laboratory. So ubiquitous is the distribution of persistent rhythmic processes throughout the living kingdom, that rhythms should probably be considered a fundamental characteristic of life, and should be added, along with such others as metabolism, growth, irritability, reproduction, etc., to the elementary-textbook definition of life. Still, in spite of the prevalence of organismic rhythmicity, there is a large proportion of biologists working today who do not realize that Claude Bernard's concept of homeostasis as a straight-line paradigm must now be modified to a rhythmic stasis. In other words, most organismic processes are not constant in constant conditions, but on the contrary continue to vary rhythmically as they did in nature, in a virtually fixed pattern over the span of a day.

With time, descriptive and comparative studies of rhythmic processes accumulated sufficient data to enable basic properties of the phenomena to be delineated. To set the stage for the rest of the book, these will be briefly outlined here and further developed as needed in the last two chapters.

One of the earliest significant findings was that the phase of a rhythm, i.e., the time in the cycle when some obvious event took place, was not necessarily restricted to a particular time of the day. For example, by placing a plant exhibiting a normal sleep-movement rhythm into a laboratory situation in which light is now offered at night and darkness during the day, the plant quickly alters the phase of its rhythm so that the leaves are in the "awake" position during the new hours of illumination. In fact, no matter what part of the 24-hour day that "light-on" is offered, the plant quickly adjusts so that the leaves are in the raised position during the times of illumination. After such a treatment, if the plant is then placed in constant conditions, the new phase relationship holds for several days, indicating that the clock-controlled rhythm has been reset by the treatment. Such phase lability is a general property of biological rhythms.

Light cycles also control the period (i.e., the time interval between two identical points in a cycle) of an organismic rhythm. The period can be increased by offering abnormally long "days" (e.g., 13.5 hours of darkness alternating with 13.5 hours of illumination) or decreased by offering abnormally short "days" (down to about 18 hours total) (Figure 1-3A and B). However, this treatment does not have a persisting effect on the rhythm; when organisms previously subjected to such unnatural day lengths are placed in constant conditions, the period of their rhythms immediately returns to about 24 hours.

Moreover, when more extreme artificial "days" are offered (e.g., 8-hour days, as seen in Figure 1-3C), the organisms "ignore" them and continue to display their natural period of about 24 hours.

In the natural habitat, organismic rhythms are strictly 24 hours in length; they are "locked" or "entrained" to this frequency by the daily light-dark cycles generated by the rotation of the earth on its axis. When plants and animals are placed in constant conditions, the rhythm persists, but the period usually becomes slightly longer or shorter than 24 hours (Figures 1-4 and 1-2). To describe this difference, Dr. Franz Halberg of the University of Minnesota described the new period as *circadian* (*circa*, about; *di* (*em*), day + -AN), i.e., *about a day* in length. Here light has yet another effect: The length of the period of a circadian rhythm is usually a function of the intensity of the constant illumination, becoming longer or shorter with increasing light intensity, depending on the species of organism being tested (Figure 1-5).

A biological clock would *a priori* be expected to be built of biochemical components, and all usual chemical reactions are known to be particularly sensitive to changes in temperature. For example, as the ambient temperature increases, atomic and molecular agitations increase so that the incidence of collisions between the atoms and molecules is augmented and their reaction rate consequently multiplied. As a rule of thumb, each $10°C$ rise in temperature causes a doubling or tripling of the reaction rate. Therefore, from a chemical standpoint, a $10°C$ rise in temperature should double or triple the rate at which a biochemical clock would run, thereby halving or thirding the length of the period of the overt rhythm. On the other hand, from a philosophical perspective no clock, living or otherwise, can be sensitive to temperature changes, for if it were it would no longer tell time, but instead become an unusual type of thermometer signaling temperature changes by its rate of running.

Therefore, when the experiments were done, it was not surprising to find that the periods of all circadian rhythms are very little affected by temperature change. A $10°C$ increase or decrease seldom causes even as large a change as 20% in the length of the period (Figure 1-6). Describing it in the argot of thermochemistry, temperature coefficients for rhythms normally range between 0.9 and 1.2—a very significant deviation from the coefficients encountered with other chemical systems, which usually produce coefficients in the range of 2.0 to 4.0.

FIGURE 1-2

Circadian sleep-movement rhythm in the bean seedling, Phaseolus, *in continuous dim illumination (signified by the open horizontal bar subtending the abscissa) and constant temperature. The parallel curved reference lines are 24 hours apart, emphasizing the fact that at this light intensity, the period of the rhythm was approximately 27 hours long [redrawn and modified from Bünning, E., and M. Tazawa (1957). Planta* **50,** *107-121].*

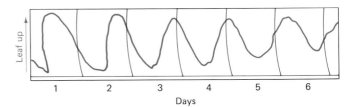

Days

FIGURE 1-3

Diagrammatic representation of the effects of light-dark cycles on the period of a typical biological rhythm. A. After many days in a 24-hour light-dark regime (of which only the last day is shown) a rhythm is subjected to two 27-hour "days" (13.5 hours of illumination alternating with 13.5 hours of darkness). The rhythm adjusts to this new regime, but only superficially; when this treatment is followed by constant darkness (indicated by shaded bar along abscissa), the rhythm instantly reverts to its natural period of about

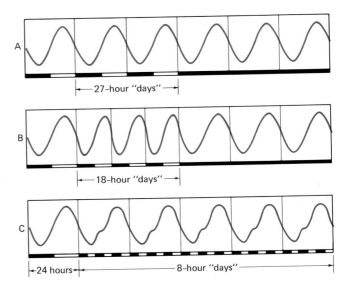

A

|←——27-hour "days" ——→|

B

|←——18-hour "days" ——→|

C

|←24 hours→|←————8-hour "days" ————→|

24 hours. B. Same rhythm subjected to three 18-hour "days." Entrainment occurs, but again the period reverts back to 24 hours when the rhythm is tested in constant darkness. Note in both A and B that the phase of the circadian rhythm in constant conditions has been set by the last exposure to light (i.e., the first peak in constant darkness comes 24 hours after the last peak in the artifical light-dark cycle). C. Same rhythm now subjected to acutely extreme "days" of 8 hours. The form of the curve is distorted by the treatment, but the period remains constant at about 24 hours.

FIGURE 1-4

Diagrammatic representation of the circadian nature of a biological rhythm in constant conditions. In the first column is shown a rhythmic function subjected to a normal daily light-dark cycle. The rhythmic pattern repeats itself each day at the same time maintaining the period at exactly 24 hours. In the second and third columns, the same rhythmic system is now shown subjected to two different sets of constant light and temperature in the laboratory. Depending on the individual experimental organism and/or the type of conditions used, the peaks may drift in one direction or the other in relation to the 24-hour day. In the first column the period is 25 hours; in the other, about 23 hours.

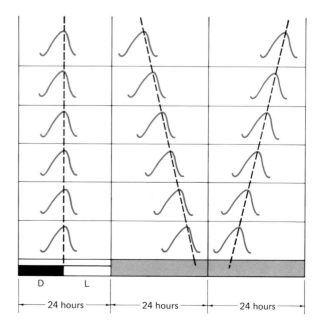

Like light cycles, temperature cycles can also be used to entrain and set the phase of biological rhythms. Figure 1-7 is a diagrammatic representation of the ability of low amplitude temperature cycles (12-hour periods of 28°C alternating with 12-hour periods of 20°C) to entrain a rhythm. As can be seen, the phase of the rhythm can be caused to synchronize with any hour of the 24-hour day. Moreover, when the organism is then transferred to constant light and temperature, the phase remains essentially unaltered, showing that the treatment has actually reset the clock-timed overt rhythm in the same manner light cycles had done.

Another property of biological clocks is their insensitivity to a great variety of chemical inhibitors, including narcotizing agents and sublethal doses of metabolic poisons. As diagramed in Figure 1-8, one would expect that treatment with a metabolic inhibitor such as sodium cyanide would certainly at least slow down any metabolic clock so that the period would become appreciably lengthened. Yet investigations have shown that while the amplitude of the rhythm is diminished with the application of such agents the period remains essentially unaltered. This insensitivity has been found to hold true even when an organism's metabolism was reduced to 5% of normal. As will be elaborated in the chapter by Professor Hastings, pulsing of many of the most common inhibitors into living systems also has little conclusive effect on their rhythms.

Another major property of biological rhythms is also one that might not be anticipated: They are innate. That is, the period is not learned, or imprinted upon organisms by the 24-hour day-night light and temperature cycles produced by the rotation of the earth. This has been demonstrated by raising animals from birth—and seeds from the time of germination—in static laboratory conditions. The developing organisms either become rhythmic *de novo*, or they are arrhythmic but can be made to become rhythmic by subjecting them to a single, *nonperiodic* stimulus.

The fruit fly provides us with a clear-cut example. In nature, the adult fruit fly (*Drosophila*) emerges from its pupal case only at dawn and this rhythm will persist in constant conditions. However, if batches of eggs are laid and made to develop in constant conditions, the resulting adults eventually emerge at all times of the day, i.e., the population is arrhythmic. It was found that even after twenty-five generations in constant conditions, when the developing larvae or pupae are given one nonperiodic stimulus—e.g., a single brief light flash, or a single abrupt change in illumination level—a rhythm was initiated in the population (Figure 1-9). Therefore, it is quite obvious that

FIGURE 1-5

The effect of different intensities of constant *illumination on the period length of circadian rhythms. The period of the mouse spontaneous locomotor rhythm increases with increasing light intensities, while the period of the activity rhythm of the chaffinch varies inversely with light intensity [redrawn and modified from Aschoff, J. (1960). Cold Spring Harbor Symp. Quant. Biol.* **25***, 11-28].*

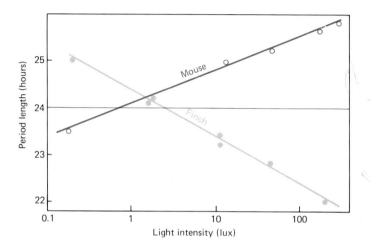

FIGURE 1-6

Diagrammatic representation of the expected and observed results of the effect of a 10°C increase in temperature on the period of a typical biological rhythm in constant conditions. When the tempera-

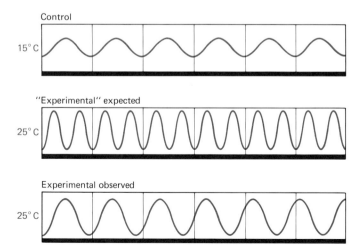

ture is increased by 10°C the biochemical clock would be expected to run at least twice as fast, generating a period of half the previous interval. However, the period is only slightly shortened by this treatment, demonstrating the virtual temperature "independence" of living clocks. Note that while the effect on the period is slight, the amplitude of the rhythm is influenced in the expected way: doubling in height.

FIGURE 1-7

Diagrammatic representation of the effect of an 8°C temperature cycle on a biological rhythm. A, B, and C represent three different, but identical, rhythmic systems maintained in constant darkness and subjected to temperature cycles each 8 hours out of phase with one another. The square wave subtending each of the rhythm curves indicates the ambient temperature cycle; the depressed portions of the square wave signify a temperature of 20°C; the plateaus, 28°C. At the time indicated by the straightening of the temperature curves the rhythms were subjected to a constant temperature of 24°C. It is seen that the phase remained unchanged, showing that the clock had been reset by the treatment.

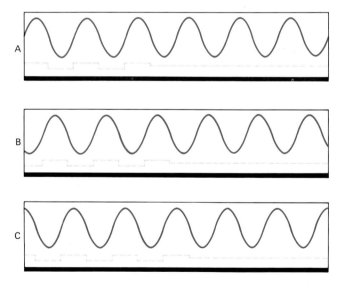

FIGURE 1-8

Diagrammatic representation of the effect of a sublethal dose of a metabolic inhibitor (sodium cyanide) on the period length of a biological rhythm. Such treatment might be expected to slow down the clockworks of a metabolic horologe, thereby *increasing the length of the period (as portrayed by the "expected" curve). When the experiment was performed it is seen that while the amplitude was reduced, the period was virtually unaffected.*

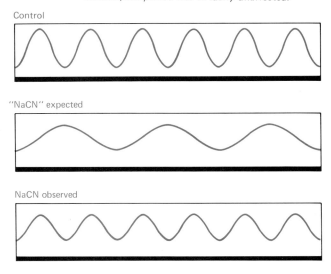

Control

"NaCN" expected

NaCN observed

FIGURE 1-9

Establishment of an eclosion rhythm in an arrhythmic population of fruit flies (Drosophila). Flies raised in constant conditions emerge randomly from their pupal cases over the 24-hour day. However, all that is needed to induce a rhythm into the population is one, nonperiodic stimulus: A light is turned on. The rhythm, thus established, persists in continuous dim illumination until all members of the population have emerged as adults [Bünning, E. (1935). Ber. Deut. Botan. Ges. **53,** *594-623].*

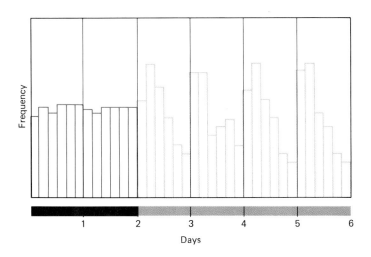

the ability to measure off periods of about 24 hours is an in-nate property of protoplasm.

At present the fundamental nature of the biological clock remains enigmatic. Using the properties of rhythms described above as model-building plinth stones, two possible kinds of clock mechanisms have been postulated by theoreticians: the external timing hypothesis, and the autonomous, endogenous timer hypothesis.

The former postulate is championed by Professor Frank A. Brown, Jr., who is one of the handful of early pioneers to attack the timing problem. His studies soon convinced him that organisms in standard laboratory constant conditions were using subtle, rhythmic geophysical forces—those that easily permeated the barriers of an experimental setup—as an informational input to time their overt rhythmic processes. It is the background work supporting the external timing hypothesis that he discusses in the following chapter of this book.

Brown's intriguing idea has spurred an active controversy, for a second camp of investigators holds a contrasting view, postulating that organisms are completely independent of the environment for their fundamental timing information, and instead are thought to possess autonomous biochemical clocks able to measure out absolute time. It was the challenge of describing this mechanism that brought Professor J. Woodland Hastings (the author of the last chapter of this book) into the field of biological rhythms. Over the past fifteen years, he and his colleagues have performed some of the best work thus far done on the biochemical nature of the elusive clock.

The remainder of the book will concentrate on a descrip-tion of the experiments leading to the development and sup-port of these two hypotheses.

HYPOTHESIS OF ENVIRONMENTAL TIMING OF THE CLOCK

FRANK A. BROWN, JR.

Morrison Professor of Biology
Department of Biology
Northwestern University

My fascination with the phenomenon of biological rhythms goes back many years to my graduate-school days. My first study in the subject was conducted just as I was finishing my doctoral program at Harvard University in 1934, when Professor John Welsh at Harvard and Professor Orlando Park at Northwestern University were reporting the phenomenon in crustaceans and insects, respectively. At that time the overwhelming majority of biologists ignored the phenomenon. Unable to accept either of the two proposed alternative explanations of the phenomenon, fully independent internal timers for these long periods or a capacity to respond to such very weak subtle, terrestrial fields as the electromagnetic, most biologists found it virtually impossible to accept the reality of the phenomenon itself. It was suspected that the investigator was simply not controlling adequately the fluctuations in some ordinary stimuli. Indeed, when I suggested to a senior faculty member whose counsel I held in high regard that I might devote my future researches to the resolution of the nature of the rhythmic mechanism, I was advised that this was no proper subject for a promising young scientist. So, dutifully, I turned to more conventional areas of biology and worked until I had achieved a full professorship, a good salary, permanent tenure—could not be fired—and then, in 1948, I returned to the rhythm problem.

The only other extensive program of study of the subject underway at the time of my return was that of Professor Erwin Bünning and his associates in Germany, who were making many valuable contributions to our knowledge of plant rhythms and their roles. For the next few years it was most gratifying to see a rapid growth in appreciation and investigation of the phenomenon in many laboratories in many countries.

Our earliest investigations of the nature of the rhythmic mechanism developed as a natural outgrowth of our ongoing studies of the hormonal regulation of color changes in the fiddler crab. Our endocrine studies were constantly beset by the fact that our crabs, though treated experimentally under the same laboratory conditions and in the same manner, would give us highly variable results from one time to another. It became, first, readily apparent that a conspicuous, deep-seated diurnal variation was occurring in the crab's color-change system. The crab's whole physiology was in good measure enslaved to what we have since come to know as a biological clock system.

Fiddler crabs inhabit the intertidal zone of the seashore. All of the plants and animals of this region are subjected to the

24-hour, day-night changes resulting from the rotation of the earth in the relation to the sun, to the ebb and flow of the tides which rise and fall with a period related to the rotation of the earth relative to the moon, to semimonthly fluctuations in the heights of the tides resulting from joint influence of both sun and moon on the tide, and to the seasonal changes as the earth makes its orbital journey around the sun. These organisms, just as their ancestors long before them, are steadily subjected to monotonously repetitious fluctuations in illumination, temperature, tidal submergence, and numerous other factors.

The crabs spend much of their lives in burrows that they dig for themselves, but as each tide ebbs and exposes the burrow entrances the crabs come milling out to feed. While the times of foraging for food vary rhythmically with the tides the crab's skin color varies with time of day. As the day dawns the crabs darken and as the sun sets they become pale.

Dr. Marguerite Webb and I learned that when the crabs are captured and brought into an experimental darkroom in the laboratory where they are deprived of all obvious clues that signal the passage of the day, they continue to darken and lighten over the 24-hour period just as their companions are doing out on the beach. This is an illustration of a 24-hour clock-timed rhythm.

Figure 2-1 is a rendition of a photograph of two crabs: one in the daytime, darkened state, the other in the characteristic nighttime light one. Before this photograph was taken, about noon one day, the light cycles had been altered for the crab on

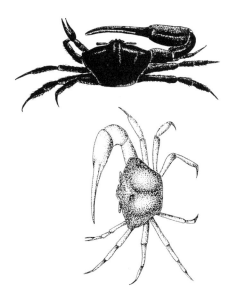

FIGURE 2-1

A drawing from a photograph of Fiddler crabs, Uca pugnax, *showing one crab in its dark, daytime phase and the other in the blanched, nighttime phase. By subjecting the right-hand crab to a reversed light cycle the phase of the color-change rhythm was reversed, enabling crabs in both phases to be photographed together.*

the right; it had received light during the night and darkness during the day for several days. This reset the phase of the crab's clock-timed, color-change rhythm. This crab now operated as if it had been collected from a point about halfway around the earth, for example, in Singapore. In the darkroom where no light changes occurred these two crabs continued to lighten and darken exactly "out of step" with one another.

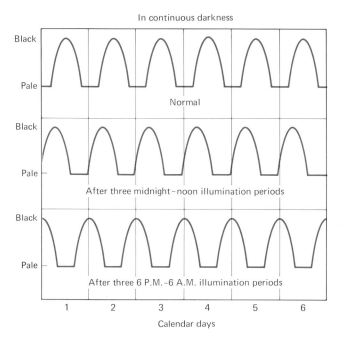

In continuous darkness

Normal

After three midnight–noon illumination periods

After three 6 P.M.–6 A.M. illumination periods

Calendar days

FIGURE 2-2

Diagram depicting resetting the fiddler crab's color-change rhythm by 3-day treatments of artificial light-dark cycles set to times different from the natural one. The new phase relations of the crabs then persist in constant darkness.

In practice, we can cause the light phase of the color-change rhythm to occur at any time during the 24-hour day by adjusting the time of day that the light period is given. Figure 2-2 is a diagrammatic illustration of such an experiment. By presenting light to crabs between midnight and noon for three consecutive days, the phases of the rhythm are moved ahead by about 6 hours; the crabs are now behaving as if we had collected them from, say, the coast of the Baltic Sea. Or, instead, if we turn the lights on at 6 P.M. and leave them on until

6 A.M. for three consecutive days, we find that the color-change rhythm is altered relative to the untreated control crabs, just as we noted for the crab in Figure 2-1. All of these new phase relationships persist when the crabs are held in continuous darkness. Quite comparable 24-hour clock-timed rhythmic systems exist in man; their existence is what makes it necessary for, and at the same time permits us to adjust our multifold daily rhythms to any other longitude of the earth as we travel by jet.

The clock-timed rhythmic system of the crabs was soon discovered to be even more complex. To disclose any existing rhythms in amount of activity "actographs" were constructed. In many biological laboratories the apparatus that is employed becomes increasingly complicated year by year as the study continues. In our laboratory, on the contrary, the apparatus often becomes simpler and simpler as we come to know more and more about what we are seeking. Figure 2-3 shows our latest model crab actograph. Shielded from light and away from the ocean tides the running activity of the crabs

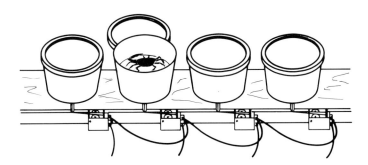

FIGURE 2-3

A simple actograph for fiddler crabs. Each light plastic container is supported by a single point-fulcrum centered on the bottom. As a crab circles within the dish the latter rocks around, closing an attached microswitch which completes a circuit causing a pen deflection on an event recorder. In this manner one obtains a complete record of the variations in the crab's running activity.

remains timed to the tides. Even if one places crabs on a windowsill so that they are exposed to the normal day-night cycle of light, the crabs' running rhythms retain the tidal periodicity. Results of such a study by Dr. Franklin Barnwell are shown in Figure 2-4. Two periods of activity occur "spontaneously" each day, about 12.4 hours apart. That the activity periods are synchronized to the tidal ones is evident from the daily progression of the tides on the crab's home beach, indicated by the dots. Here, then, despite the fact that the crabs are in a 24-hour light-dark regime and that the color-change rhythm is operating on a 24-hour basis, the running

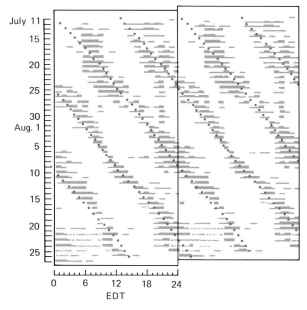

FIGURE 2-4

*The black bands depict the changing times of day of spontaneous locomotor activity of a fiddler crab (*Uca minax*) during 45 days in natural illumination in the laboratory. The lines of dots mark times of high tide on the crab's home beach. The record is repeated and displaced upward one day to facilitate the reviewing of the passage of the patterns of activity from one day to the next [Barnwell, F. (1966).* Biol. Bull. **130**, *1-17].*

activity of the crabs comes, on the average, 50 minutes later each day, just as rise the tides and the moon, whose gravitational pull the tides obey. Clearly the crabs are able to time at one and the same time both solar-day and lunar-day activities.

Figure 2-5 shows the activity pattern obtained over a month of another fiddler crab, a Costa Rican one, held away from the

FIGURE 2-5

*The spontaneous locomotor activity (black marks) of a Costa Rican fiddler crab (*Uca princeps*) during a month's sojourn in continuous darkness. The open circles mark the times of high tide on the crab's home beach. Note the two bursts of greater activity (arrows) which occur at the same time of day with a semimonthly interval. The record is repeated and displaced upward one day to facilitate the viewing of the passage of the patterns of activity from one day to the next (unpublished record by Franklin H. Barnwell, University of Chicago).*

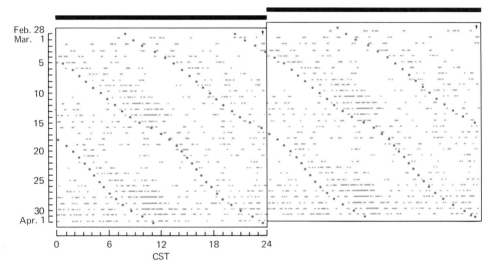

FIGURE 2-6

There are two hypotheses concerning the nature of the primary timer of the biological clock system: (1) Each organism individually is by its physical nature a timer independently capable of measuring closely the durations of days, tides, months, and years; (2) all organisms depend steadily on an inflow of pervasive, subtle information from their rhythmic physical environment for the timing of their rhythms having close to these same periods.

tides in constant darkness. Passage of the tides over the day on the crab's home beach is indicated by the small circles. Inspection of Figure 2-5 shows that at half-monthly intervals there is increased activity during the late morning hours. Here, then, is a remarkably precise, persistent, semimonthly rhythm in running activity. Such semimonthly rhythms are an expected consequence of additive actions of a solar-day rhythm and bimodal lunar-day or tidal rhythm. By periodic mutual reinforcement ("beats") the longer cycles are produced. Many biological processes are known to possess closely this frequency as, for example, reproductive cycles of many marine animals and plants whose breeding times, once established for any given area, can often be predicted with great accuracy thereafter simply in terms of phase of the moon, time of day, and times of year. Even the human female possesses a semimonthly rhythm in that the period from menstruation to ovulation, or ovulation to menstruation, averages close to a semimonth. The latter suggests, therefore, that even the human possesses simultaneously, both solar-day and lunar-day clocks. Indeed, studies in a number of other laboratories, including those of Dr. John Palmer at New York University and Dr. E. Naylor at the University of Swansea, Wales, have given us reason to postulate that a biological clock system for timing lunar-tidal periods may be as widespread among living creatures as the solar-day timing system.

What is the nature of the biological clock timing mechanism? Figure 2-6 illustrates the two presently popular alternative postulates: the internal timer and the external timer hypotheses. The *internal timer hypothesis* is derived as follows. In nature, organisms are rhythmic with periods that are of the same length as the natural ocean tides, days, months, and years. When one experimentally deprives the organism of all obvious environmental clues which usually reflect these cycles and yet finds that the rhythms persist, it is concluded that the organisms must have some internal, biochemical timer that independently measures these periods. And, of course, this may be readily rationalized. After all, living things have been evolving for millions of years in a rhythmic environment and so it is perhaps not surprising that they might have generated, as an adaptive feature for predicting the occurrence of favorable and unfavorable times, the ability to measure these periods on their own. Contrasting with this view is the *external timer hypothesis.* Proponents of this hypothesis are agnostics who question whether the conditions are truly constant for the organism. They are aware that many parameters of the physical environment that undergo daily, tidal, monthly, and annual variations are still pervading the "constant" conditions as well as the organisms themselves. Were organisms sensitive to the fluctuations then they might be using information to time their observed rhythms displaying the same, or close to the same, periods. Thus, advocates of this hypothesis postulate that the organism's biological clock comprises a capacity to receive such timing information from the environment and transduce it into the observable biological rhythms. At present there is insufficient evidence to exclude either hypothesis. Perhaps both kinds of timers are available to organisms.

Figure 2-7 is an illustration of a common phenomenon that has been offered for many years as proof that the timer is an independent internal one. These are records, obtained by Dr. Patricia DeCoursey at the University of Wisconsin, of the running activity of two flying squirrels, one for 26 days and the other for 23 days, both kept in constant darkness. Under these conditions, one squirrel began running about 2 minutes earlier each day and therefore was generating a period of about 23 hours and 58 minutes. The other began running about 21 minutes later each day and therefore displayed a period of about 24 hours and 21 minutes. If one had a dozen or more flying squirrels under the same conditions, each would probably have a period that was a little different from that of every other one. And so the argument ran that these differences obviously proved the existence of an independent internal timer inside each individual because such a variety of natural geophysical periods close to 24 hours would not exist, nor, even if they did exist, would every different flying squirrel be expected to pick on a different environmental period to time its individual rhythm. This presumed proof for internal timing depended upon the arbitrary assumption that any observed circadian rhythm must be being timed by an underlying circadian timer displaying exactly the same circadian period. Under these assumed conditions the rhythms were termed "free-running" with their period directly determined by the postulated circadian clock.

However, the occurrence of circadian periods differing from 24 hours is no proof whatsoever that the rhythm timer is also circadian and within the organism. It is both theoretically and even demonstrably possible for an organism to generate a regular period other than 24 hours using a clock which runs at 24 hours. If I were to say to the reader that if he would open this book every 24 hours and 21 minutes (i.e., with a circadian period like one of the squirrels) for the next five days, I would

FIGURE 2-7

The spontaneous locomotor activity of two flying squirrels (Glaucomys volans*), as recorded in running wheels in constant darkness. The period of one squirrel's rhythm averaged about 23 hours and 58 minutes; that of the other's averaged 24 hours and 21 minutes [DeCoursey, P. (1961). Z. Vergleich. Physiol.* **44,** *331-354]*.

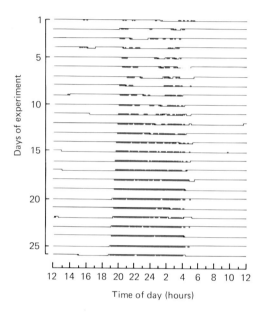

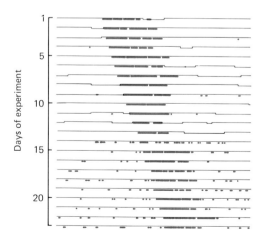

Time of day (hours)

give him a million dollars, I am sure he would manage. For five days all he would worry about was keeping his trusty wristwatch wound so that he could arrange to open the book 21 minutes later each day. This simple analogy also suggests a general hypothesis which could account for the timing of any circadian period. It is theoretically possible for an organism to use a 24-hour timer, or even a lunar-day one, to time circadian rhythms of diverse frequencies. In other words, frequency transformations may be the explanation of the often-observed deviation in period of a clock-timed rhythm from that of a natural geophysical one. The period of the clock may differ from that of the rhythm it times. The postulated means by which an overt rhythm having a period different from 24 hours is timed by a 24-hour timer has been termed *autophasing*.

As described in Dr. Palmer's chapter, changing an organism's temperature even over a wide range changes a circadian period only very slightly. Also, the period of a rhythm may similarly be altered very slightly by changing the intensity of the unvarying illumination to a different level. Proponents of the internal timer hypothesis postulate that these changed periods reflect an alteration in the speed at which the circadian clock is running. On the other hand, in terms of the extrinsic timer hypothesis that I am going to develop in this chapter, the basic timer remains a 24-hour, or a 24.8 hour, one with the observed very slight temperature and light dependency of the period being a consequence of action of these factors on the frequency-transformation, or autophasing, mechanism by which the diverse circadian periods are generated from the basic clock system.

For the first five or six years that my associates and I worked with biological rhythms, our working hypothesis was the prevailing one, namely, that every organism had its own autonomous internal timer. However, working with Dr. Ross Stevenson, now at Kent State University in Ohio, we obtained some results which planted the first seeds of doubt. We had a batch of fifteen oysters shipped from New Haven in Connecticut to Evanston, Illinois, where we studied their shell-opening activity in the laboratory in glass vessels in continuous low illumination. In New Haven the oysters were presumed to open their shells maximally during the times of high tide and close them when exposed during low tide. We studied the oysters' activity in the laboratory by tying one end of a thread to their upper shell and the other end to lever-operated recording pens so that we could record continuously and automatically the tendency of the oysters to open, in order to learn how it might

vary with the time of lunar-day (or tide). Analyzing the data
of the first 2 weeks in Evanston we discovered that the oysters
opened their shells most at the times of high tide in their
former home waters, about a thousand miles to the east
(Figure 2-8).

FIGURE 2-8

*The average number of minutes of open shells of
fifteen oysters transported from New Haven, Connecticut, to Evanston, Illinois, and maintained in
constant conditions. The average number of minutes open for hours of the lunar-day, the horizontal axis, is plotted. A: For the first 2 weeks in
Evanston (arrows indicate time of high tide in
oysters' home waters); B: for the next 2 weeks;
C: for the last 2 weeks of the study. The time of
maximum shell opening is seen gradually to shift
to, and then stabilize at, the times of upper and
lower transits of the moon [Brown, F. A. (1954).
J. Physiol. 178, 510-514].*

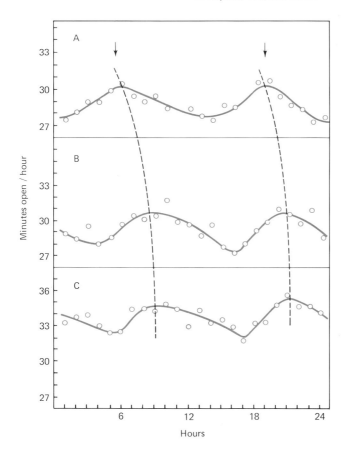

On viewing the comparable results for the second 2 weeks, we observed a surprising change; the peaks of the rhythm had moved to times 3 hours later in the lunar day. Similarly unexpectedly, the new phase relationship appeared to remain unchanged for the third 2 week period in the laboratory. After a month or two of pondering these results, it occurred to me to look at the almanac, and there I discovered that the peaks had shifted to the times of upper and lower transit of the moon in Evanston, Illinois. It is during these two times each day that theoretically it would have been high tide in Evanston, if Evanston had been a seacoast town and had had a tide. Since the oysters were strangers to Evanston, these results at first seemed thoroughly absurd until it was recalled that there are tides in the atmosphere which have lunar components as well as solar ones, just like the tides of the oceans. One way these tides are apparent at ground level is in the fluctuations in barometric pressure. If an organism somehow possessed the capacity to resolve these pressure rhythms then these might be used to time the organism's tidal rhythms as well as set them to local time as we observed here. However, we found very shortly, as will become evident later, that the oysters were probably not responding directly to the rhythmic pressure changes in the atmosphere.

There are many other parameters in the environment that change in manners correlated with the lunar-day and solar-day atmospheric tides. One is gravity; others are geomagnetism, the electrostatic field, and background radiation. In our laboratory in the last ten years we have been able to demonstrate by simple, conventional kinds of experiments that organisms are extraordinarily sensitive to diverse parameters of electromagnetic fields at intensities normally experienced in nature. Indeed, the evidence indicates that their receptive mechanisms are specially adapted to these very weak fields. It could perhaps be one or more parameters of these fields that provides the information to organisms to enable timing of their rhythms.

Figure 2-9 depicts diagrammatically the general thesis I shall develop in the remainder of this chapter, namely, that organisms are duplex systems with respect to their rhythmicity. One may think of any plant or animal as possessing one rhythmic system underlying another one. In Figure 2-9 the outer ring represents the overt physiological rhythms that we may observe in the organisms in nature and in the laboratory. These are adjustable recurring patterns that can be altered by such factors as changed times of day of 24-hour light-dark cycles, differing photoperiods, temperature changes, times of feeding, and training, thus allowing the organism to adapt to its specifically detailed habitat conditions. Examples were seen in the fiddler crab whose color-change rhythm could be adjusted to any longitude on the surface of the earth simply by regulating the times that lights are turned on and off, or whose activity

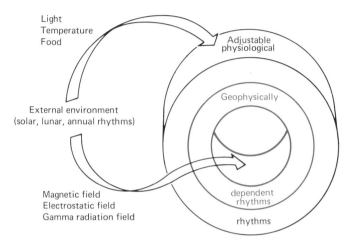

The rhythmic organism

FIGURE 2-9

Diagrammatic representation of the duplex nature of biological rhythms. One, represented by the outer ring, includes the circadian, tidal, and other rhythmic patterns adapting the organism to the specific rhythmic patterns of its environment. The other, represented by the inner ring, describes underlying, deep-seated, rhythmic fluctuations, direct responses to subtle, rhythmic forces of the physical environment [Brown, F. A., Jr. (1964). Advan. Astronaut. Sci. **17**, *29-39]*.

could be adjusted to the specific tidal times of a local beach by a period of sojourn on that beach. Underlying such adjustable physiological rhythms are deep-seated geophysically dependent ones represented by the inner ring in Figure 2-9. These geophysically dependent rhythms cannot be altered; they are locked to the fluctuations in the earth's geophysical fields to which they are direct, fundamental biological responses. We have reason to believe that the geomagnetic and geoelectrostatic fields, and background radiation are the effective forces. It will be obvious to the reader that I am going to develop the

thesis that these forces provide the fundamental timing informational inflow for the organism in what has hitherto been
deemed to be constant conditions.

Professor Erwin Bünning, a number of years ago, reported a
very interesting phenomenon suggestive of just such a duplex
rhythmic system. He studied the "sleep movements" of bean-
seedling leaves (Figures 1-1 and 1-2) and, after maintaining one
of his plants in constant illumination and temperature, Professor Bünning illuminated it with a brighter light for a few
minutes (Figure 2-10) as the bean entered its nighttime phase.
The bean plant seemed to interpret this light as signaling the

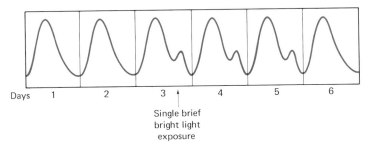

Days   1   2   3   4   5   6

                         Single brief
                         bright light
                         exposure

FIGURE 2-10

*Tracing of a bean-seedling sleep-movement rhythm.
With the exception of a brief exposure to a bright
light on day 3, the plant was maintained in constant dim illumination and constant temperature.
Note the leaf response to the bright light and the
fact that this "response" recurred each day thereafter without any further stimulus [redrawn from
Bünning, E. (1956). Naturw. Rundschau 9, 351-
357].*

onset of a new day and started to raise its leaves. When this
light was turned off the bean seemed to say "false alarm" and
the leaf resumed its previous sleep rhythm. Now, strikingly,
every succeeding day, at close to the same time, the leaf was
again raised briefly without any further stimulation. This re-
petitive action seems analogous to a system in which a magnetic
recording tape is recycling once about every 24 hours and
upon which a new entry has just been inscribed. At the same
time each day the entry is replayed. There are many other
rhythmic phenomena that seem to reflect a similar behavior.
For example, a bee can be trained to fly to a sugar-water feeding station at any given time of day, e.g., at 9 in the morning
or 1 in the afternoon, or even both times. Once the time is

FIGURE 2-11

*Recording respirometer-barostat assembly. As orga-
nisms consume oxygen from air in the flask, O_2 is
replaced from the collapsible reservoir and CO_2 and
NH_3 absorbed by KOH and cupric chloride. The
weight of the "diver" increases (buoyancy decreases) as O_2 is consumed with the increase recorded on a rotating drum. The diver may be maintained in a hermetically sealed system in constancy
of all ordinarily controlled variables as well as of O_2
and CO_2 and pressure* [Brown, F. A., Jr. (1960).
Cold Spring Harbor Symp. Quant. Biol. **25**, 57-71].

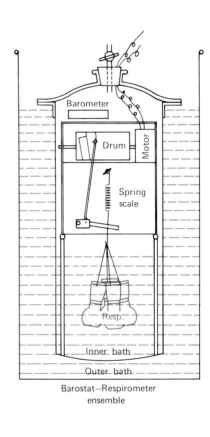

Barostat—Respirometer
ensemble

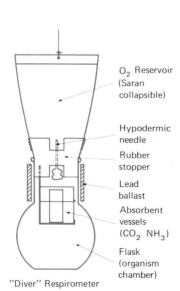

"Diver" Respirometer

learned the bee will reappear at the feeding station at the learned time for several days even after the food is no longer provided. Indeed the gross shifting of a whole daily pattern of behavior in response to move to a new time belt may not be really a fundamentally different phenomenon from these.

In order to investigate in greater depth the nature of any external timing force, or what some of the characteristics of its influences might be, we invented an automatic recording respirometer for measuring the metabolic rates of living things (Figure 2-11). Metabolic rate was chosen since now we would be dealing with one of the very basic phenomena of life itself. The respirometer was designed to permit us to study the organisms not only in constancy of all the factors physiologists normally control, but pressure as well. One organism we studied extensively was the potato. This organism seemed especially well suited for long-term studies of metabolism, for the sprouting eyes are metabolically quite active and the surrounding tissues rich in stored nutrients. The potatoes, too, seemed to possess no significant, overt, obscuring circadian rhythm. We cut eye-bearing, cylindrical plugs (Figure 2-12) from the pota-

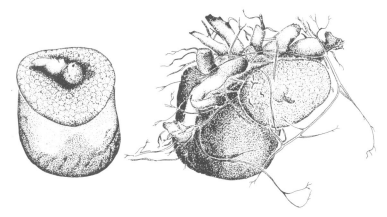

FIGURE 2-12

Potato plugs used in metabolic studies. The plug on the right had been sealed in a recording respirometer for more than 5 months; it sprouted an abortive rhizome system complete with a "new potato" growing from it.

toes, placed them in the respirometer vessels, and hermetically sealed them within barostats where all such factors as light intensity, temperature, O_2 and CO_2 tensions, humidity, and ambient pressure were without any periodic fluctuations. Figure 2-13, which is one that I think should become a classic if

FIGURE 2-13

Results of a ten-year study of the mean daily respiratory patterns in the potato. The left-hand column shows the patterns for each of ten consecutive years plotted as the average for each year. Notice the three recurring peaks each day: 7 A.M., noon, and 6 P.M. In the right-hand column the same data are used to compute the mean daily curves for all ten Januarys, Februarys, etc. Note the annual modulation in form and amplitude of these curves [Brown, F. A., Jr. (1968). Scientia (Milan) **103**, 1-16].

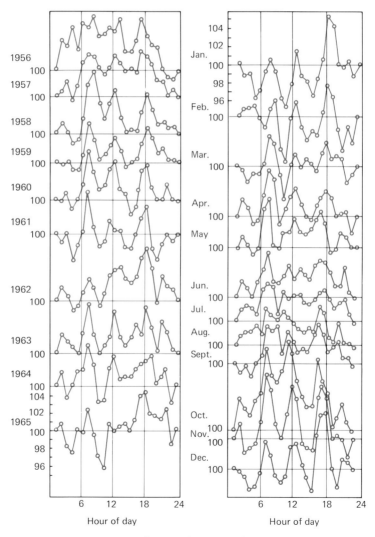

Potato O_2 consumption

only for the massive amount of data that has gone into it,
describes the results of the study.

Represented in Figure 2-13 are ten years of hourly data,
about a million and a half "potato-hours," and nearly ninety
thousand consecutive calendar hours of data. From inspection
of the left-hand column of mean daily patterns of metabolic
fluctuations for each of the ten years it is obvious that about
sunrise each morning the rate of respiration increases rapidly to
a peak at about 7 A.M., is followed by a mid-morning fall, followed by another peak at the noon hour, a mid-afternoon fall,
and another peak at 6 P.M. Then as the sun sets the oxygen
consumption decreases to a nighttime low. Although the cycles
are distorted differently between one year and another, the
three peaks come in on schedule. One year, 1962—and I shall
refer to this later—a fourth peak made its appearance in the
early evening hours and gradually faded over the next three
years. What can possibly be the explanation of not only the
three recurring peaks but also of the distortions and additional
peaks during these years? One might think intuitively that
since relative movements of the earth and sun are so regular
that any yearly patterns should be nearly identical. However,
the sun's field in which the earth is rotating is steadily changing
its pattern as activities within the sun itself fluctuate. One
would, therefore, not expect the form of any diurnal pattern of
response, if actually geophysically dependent, to be exactly the
same year after year. Indeed, these very differences argue for
external, subtle, geophysical control.

When the same data are treated in a different way, i.e., by
computing the mean daily curves for all ten years for each
month separately (Figure 2-13, right-hand column), it becomes
apparent that there is a major annual modulation in the form
and amplitude of the daily cycles. In January, the pattern
shows relatively low values in the morning and high ones in the
afternoon. As we go through the year, in summer the noon
peak becomes less conspicuous, and by the time we get to
October the overall amplitude of the daily cycle is very high
with the morning peak about as high as the noon and evening
peaks. In spite of the large annual change in gross characters of
form and amplitude of the daily curve, the potato tells you, in
effect, that it steadily "knows" essentially not only the hour of
the day but also the season of the year. Its metabolic pattern
varies systematically with the celestial longitude of the earth as
it makes its annual journey around the sun.

Figure 2-14 illustrates a number of similar daily patterns in
some organisms in addition to potatoes. The mean daily cycle

for the potato for the first three years of our study is plotted
for reference. Below the potato cycle is the mean daily respiratory pattern of carrot slices during an 8-month study. Note
that the times of the peaks are identical to those of the potato,
as are also the three peaks in the respiratory rates of mealworm
larvae which were reared in a crock of rolled oats and studied
for 9 months by Dr. Bonnalie Campbell (now at Baylor University), and in the metabolic rate in sprouting bean seeds
measured by Dr. Edward Lutsch (now at Northeastern Illinois
State College). In the case of the bean seedlings—monitored
continuously for three years—the seeds had been stored in a
darkroom for three years prior to the initiation of the study.
This means that by the end of the study the last beans used
had not experienced a single day-night cycle of light or temperature for more than six years and yet they still showed the
distinct mean daily pattern for each of the three years. The
topmost daily pattern was not a record obtained by using a
respirometer but is instead the mean daily modulation in the
locomotor activity pattern of mice in actographs as a response
to a small increase in gamma radiation. The method of recording the influence of this subtle factor on activity is shown in
Figure 2-15. For 3-day durations a very weak ^{137}Cs gamma
source was put just under the table between the cages at one
end of the table. Every third day at the same hour the gamma
source was shifted to the other end of the table. This schedule
of alternation continued for 3 months. By this procedure two
of the four mice were exposed to a gamma field (about 8
X background) five times higher than the other two mice (1½
X background) which served as controls. This increase in background radiation is well within the natural range for the field
varies more than this as one travels over the earth. For example, going from Evanston, Illinois, to Woods Hole, Massachusetts, on Cape Cod, the strength of the natural background is
doubled. In places in South America it is reportedly fifty times
higher, and in a few scattered places in Asia as much as a hun-
dred times higher than in Evanston. So, when one asks the
mice whether they "feel" a fivefold change in background
radiation it is a natural question. The method used permits one
to compare the activity of all four mice, in all four actographs,
at all positions on the table with respect only to the single variable, radiation intensity. The results are given in Figure 2-16.
Points falling above the zero line indicate that the activity at
this time of day was depressed below control values; points
falling below the zero line indicate increase in activity above
control values. Plotted in the same figure are the fluctuations
in metabolic rate of potatoes during the same year in respi-

FIGURE 2-14

*Mean daily metabolic patterns in sprouting pota-
toes, carrot slices, mealworm larvae, and germina-
ting beans, and the pattern of a daily variation in
the response of mice to a fivefold increase in background radiation [Brown, F. A., Jr. (1969). Can. J.
Botany 47, 287-298].*

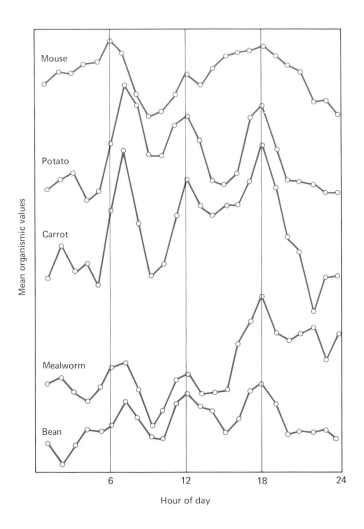

Hour of day

FIGURE 2-15

*Actograph used in study of response of white mice
to a slight increase in background radiation. The
cage is supported in the center by a single axial
fulcrum. Rocking of the cage closes a micro-switch,
an event registered on a recorder. Food and water
were provided* ad libitum *in the center of each
actograph. Radiation sources (^{137}Cs) are placed*

*beneath the mice, under the table, and are alter-
nated between two pairs of mice at 3-day intervals.*

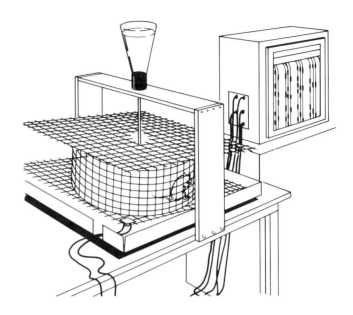

FIGURE 2-16

*The mean daily pattern of variation in effect of a
fivefold gamma-field increase on spontaneous
activity of mice in natural illumination in the laboratory as compared to a simultaneously recorded
mean daily respiration curve for potatoes sealed in
respirometers. Whether the increased radiation increases the amount of activity or decreases it is
seen to depend on hour of the day. More than 4000
mouse-hours of data and 100,000 potato-hours of
data contributed to these curves [Brown, F. A., Jr.,
Y. H. Park, and J. R. Zeno (1966). Nature 211,
830-833].*

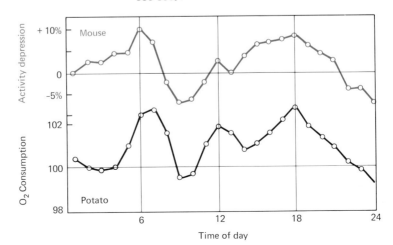

rometers in another part of the laboratory building. Comparison of the form of this daily variation in response of the mice to the form of the cycle obtained for the potato, and also that of the carrot, mealworm, and bean suggests that the mechanism governing the strength and sign of response of the mouse to the increased gamma fields is probably the same as the mechanism involved in these other organisms in regulating their daily fluctuations in rates of oxygen consumption.

Another very instructive study was that of Dr. Leland Johnson, now at Augustana College, South Dakota, who investigated the respiratory rates of chick embryos during the early stages of their development—long before they hatched. The chicks were placed, while still in their shells, in recording respiromaters, a modification of the kind shown in Figure 2-11. Results are summarized in Figure 2-17. At 4 days of incubation the chicks exhibited a daily metabolic pattern which in the spring (the dotted curve) resembled in a striking manner that reported for potatoes, beans, and several other organisms. Peaks in metabolic rate occurred about 7 A.M., noon, and 5 P.M. These studies were conducted during 1962 and 1963, the year that the potatoes (Figure 2-13) displayed an additional evening peak, and here we note that the same was evident for the chick diurnal pattern. The diurnal cycle for these 4-day chicks for fall and winter, the dashed and solid curves, respectively, displayed generally similar diurnal patterns except for now having a 6 to 8 A.M. minimum instead of a maximum at this time. The situation was much the same for the chicks during their fifth day of incubation except for an apparent inversion of the anticipated noontime maximum in the springtime. At 6 days of incubation, except for a very early morning maximum for all three seasons much like that occurring for the 5-day embryos, there seemed to be essentially no common pattern for the day.

By day 7, however, the daily pattern in oxygen consumption changed radically. As the sun rose the oxygen consumption of the developing chick increased, remained high all day, and then dropped as the sun set. The same pattern recurred on day 8. It is important to remember that these chicks had never seen a sunrise nor sunset, nor the equivalent light change, nor had they felt the additional warmth of the daytime, and yet, at this time, their metabolic pattern came essentially to match that of normal adult chickens. The embryologist, Professor Victor Hamburger, with his associate, Dr. M. Balaban, was discovering at the time of Johnson's work that at 4 and 5 days of incubation chicks have no functional eyes, functional muscles, nor functional nerves. In other words, they are essen-

FIGURE 2-17

Mean daily metabolic patterns of chick embryos during early development in what biologists customarily consider to be constant conditions. For 4 to 6 days of incubation the dotted curves are for spring, the dashed for fall, and the solid for winter. Of the bottom two patterns, both for spring, the solid curve is for day 7, the dotted for day 8. All are 3-hour moving means [calculated from original data of Johnson, L. G. (1966). Biol. Bull. **131;** *308-322]* .

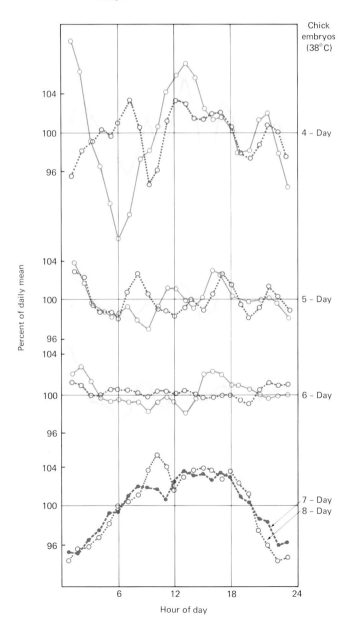

tially living "vegetables." By 6 days of incubation these struc-
tures become functional as isolated systems, but it is not until
7 days of incubation that the individual systems become inte-
grated and coordinated responses may be elicited. It seems
quite significant that simultaneously with this latter stage of
development a rhythmic pattern of metabolism comparable to
that of an adult also develops. By day 7 of incubation the
chick, which is an animal highly adapted for daytime activity,
is able to set its activity pattern adaptively to the light-dark
cycle of the particular time zone without ever experiencing a
single day-night cycle in nature. The only plausible explanation is that timing information in terms of some subtle geophysical field is available to the chick and the bird is able to
employ it in adaptively phasing a circadian pattern.

Now, if all the foregoing were not enough to lead one to
conclude that timing information must be steadily available
from the outside even in hitherto presumed constant conditions, Figure 2-18 depicts a discovery we made more than a
decade ago. In the upper left-hand corner of this figure is diagramed the form of the mean solar-day tide of the atmosphere.
The pressure rises to a high at 9 to 10 in the morning throughout the year and falls to a low at 7 or 8 in the evening during
the month of June. As the photoperiods shorten with the seasons, the evening low gradually moves to about 2 P.M. in
December. We see then that this solar-day tide of the atmosphere possesses a conspicuous annual modulation.

Now, this little solar-day tide of the atmosphere, like a little
rowboat on a very rough ocean, rides up and down the weather-associated pressure patterns as they move over the conti-
nent so that it is tipped one way and another and variously distorted. Similarly, the forms of the day-by-day metabolic fluctuations of the potato are also distorted. The height of the
morning peak on any given day is correlated with high statis-
tistical significance with the rate of barometric pressure change
between 2 and 6 that same morning. The sign of the correla-
tion is positive in the spring and summer and negative in the
late fall and winter. The height of the 6 P.M. peak is negatively
correlated all year round with the barometric pressure change
between 2 and 6 that same afternoon. It should be empha-
sized that the potatoes are hermetically sealed in constant
ambient pressure, and therefore are not exposed directly to
these external pressure changes. Despite this, the daily changes
in amplitude at the times of these two peaks tell us day by day,
far beyond chance, how the atmospheric tide is being distorted
by the weather fronts. An organism might inherit clocks but it
is beyond plausibility that it can inherit a program of all the

weather changes that are to occur during its lifetime.

The noontime peak also varies from day to day in amplitude, but this bears no relationship to barometric pressure
changes. We searched for 6 months before we found that the
vagaries of this peak were telling us with high statistical significance what the mean outdoor temperature was that same day
as it was being recorded by the U. S. Weather Bureau. The
fluctuations in the amplitude of this peak were telling us in
the warmth of the laboratory, for example, whether the average outdoor temperature was 50° above or was 5° below
zero Fahrenheit for that day. The potato was doing this
despite the fact that it was maintained at a constant temperature which did not vary more than a few thousandths of a
degree (even this was ascribable to the cycling in the thermo-
static controls). Therefore all three peaks reflect in significant
measure events going on in the external environment in spite
of the fact that the organisms are maintained under conditions
held more rigorously constant than those with which most
physiologists normally work. Again we view here very strong
evidence that information other than of a kind we are screen-
ing out is steadily getting to the organisms and they are able
to respond to it.

Another organism we studied was a mammal—the hamster.
In 1964 we put two male hamsters on a running wheel for
about half the year during a 12-month study in a light regime
consisting of darkness between 6 P.M. and 6 A.M. alternating
with 12 hours of light. At the end of a year we simply averaged
the amount of running activity for each hamster for each hour
of the day and constructed the upper left-hand curve portrayed in Figure 2-19. It is seen that the animals confined their
activity largely to the times of darkness. However, they actually anticipated the onset of darkness in that their hourly activity rate was already quite high by the time of light-off. Maximum activity occurred 1 to 2 hours after dark onset and then
the running activity decreased throughout the dark period to a
point where it had practically ceased by the time of light-on.
At the end of the year these two hamsters were replaced by
two other males for the next 12 months of recording. As seen
in the lower left-hand curve of Figure 2-19, the results were
essentially the same. It is therefore quite clear that a cycle of
12 hours of light alternating with 12 hours of darkness pro-
duces in such hamsters a more or less characteristic 24-hour
pattern of running. The following question was now asked: Is
there a lunar-day component present in these data? If one
existed it would be expected to possess peaks and troughs moving across these daily cycles at the rate of 50 minutes a day, or

FIGURE 2-18

*Diagram illustrating the specific times of day for the potato (*Solanum*) when specific metabolic deviations correlated with concurrent daily parameters of barometric pressure changes and of temperature levels. These continued for the three year study even though the potatoes were shielded from all direct influences by these two factors. [Brown, F. A., Jr. (1962). Ann. N. Y. Acad. Sci. 98, 775-787].*

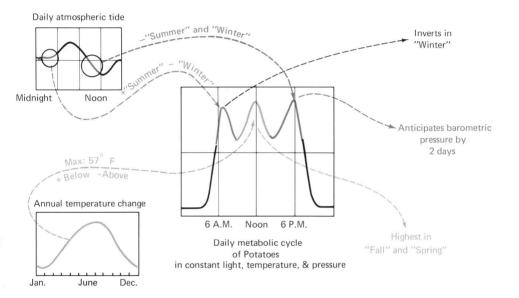

Daily atmospheric tide

"Summer" and "Winter"

Inverts in "Winter"

Midnight   Noon

"Summer" — "Winter"

Anticipates barometric pressure by 2 days

Max: 57° F
+ Below −Above

Annual temperature change

6 A.M.   Noon   6 P.M.

Daily metabolic cycle
of Potatoes
in constant light, temperature, & pressure

Jan.   June   Dec.

Highest in
"Fall" and "Spring"

FIGURE 2-19

Daily and monthly variations in activity of hamsters maintained in alternating 12-hour periods of light and darkness. The solid black horizontal bars signify the times of darkness. On the left are the mean daily activity patterns for two hamsters from 1964-1965 and of two different hamsters during

*the next 12 months. On the right are the mean
daily amounts of activity as they vary with day of
the synodic month for each of the two years
[Brown, F. A., Jr. and Y. H. Park (1967). Proc. Soc.
Exper. Biol. Med. 125, 712-715].*

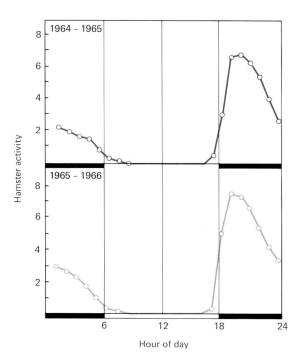

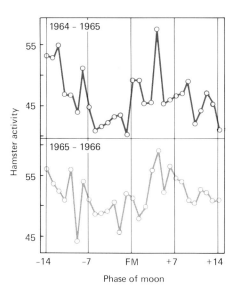

exactly once a month. Its peak or peaks would be expected to supplement the 24-hour one. By averaging total daily activity and plotting it with respect to the phase of the moon for each year separately, we derived the curves shown on the right side of Figure 2-19. The mean daily activity for each year was found to be lowest during a few days before full moon and new moon and then highest 4 days after full moon and during the days just following new moon. Comparing the highest values with the lowest, one finds about a 40% difference. Even

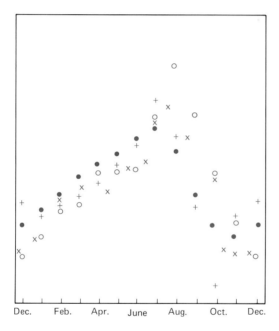

FIGURE 2-20

Annual variations in potato (●) and bean (o) metab-olism, nitrate reduction in an alga (x), and orienta-tional response of flatworms (+) to a very weak radiation source [assembled from Brown, F. A., Jr. (1968). Scientia **103**, *245; Lutsch, E. F. (1962). Ph.D. Dissertation; Kessler, E. and F. C. Czygan (1962). Experientia* **19**; *89; Brown, F. A., Jr. and Y. H. Park (1964). Nature* **202**, *469].*

if one smooths out the curves by a 3-day moving mean there is still for the two years a synchronized (about 15%) variation in activity of the hamsters maintained in the presumably constant conditions other than alternating 12-hour periods of light and darkness. This finding reemphasizes the fact that some form of information that fluctuates in a monthly pattern must be

reaching the hamsters. Highly suggestive of a specifically detailed pattern of whatever are the effective geophysical parameters is the relatively detailed parallel patterns for the two
years for the hamsters. On the basis of observations such as
these, we believe that the primary organismic clock system is
not a circadian, but rather a solunar one. We shall see further
evidence for this later.

In every case with which I am familiar where an animal or
plant has been carefully and quantitatively investigated in constant conditions over the course of more than a year, there has
been found to be a clock-timed annual variation in the process
studied. Figure 2-20 shows such clock-timed annual variations
reported for potato and bean metabolic rates, nitrate reduction
by an alga, and the orientational responses of flatworms to a
twofold increase in background radiation presented on their
right. The beans used in establishing this curve are the same
ones referred to in Figure 2-14; they had not been exposed to a
day-night cycle of light or temperature for three to six years
prior to this study. I could cite other cases: Dr. David Davis, of
North Carolina State University, followed the food intake and
gain in weight of woodchucks maintained in constant conditions for nearly three years. He found an annual rhythm with
maximum food consumption in July and August and minimum
in late fall and winter. These times could not be altered by
experimental changes in light or temperature. It seems significant that annual rhythms in such different physiological processes all reach their maxima in late summer and minima in
late fall in organisms held in conditions that are usually deemed
to be constant for them.

Another kind of observation not only confirmed impres-
sively the absence of constant conditions in all previous laboratory studies, but at the same time provided a very significant
clue as to what might be one of the effective subtle factors.
Over more than fifteen years we have throughout each summer
measured the metabolic fluctuations of fiddler crabs by means
of our automatic recording respirometers. The usual results are
illustrated in Figure 2-21, where the records for two summers,
1956 and 1957, are shown. The typical reproducibility of the
daily patterns of metabolic fluctuation for corresponding days
of a natural semimonth between two years is shown on the
left. These semimonthly recurring patterns are explicable almost wholly in terms of the concurrent presence of a bimodal
lunar-day component and a unimodal solar-day one, both of
relatively large mean amplitude. These components are shown
in the same figure, on the right. The day-to-day distortions of
the solar-day component were, even in the constant-pressure

FIGURE 2-21

On the left are shown the diurnal patterns of fiddler-crab metabolism in constant conditions showing how they change their form systematically with day of the semimonth and how these patterns are reproducible from one year to another. This semimonthly recurrence of daily patterns results from the simultaneous existence in the crabs of the lunar-day and solar-day rhythmic components illustrated on the right. Also illustrated is the mean form of the solar-day tide of the atmosphere for the month of June, with thickened portions marking the times of day-to-day correlations between crab metabolism and barometric pressure [from Brown, F. A., Jr. (1960). Cold Spring Harbor Symp. Quant. Biol. **25**, 57-71].

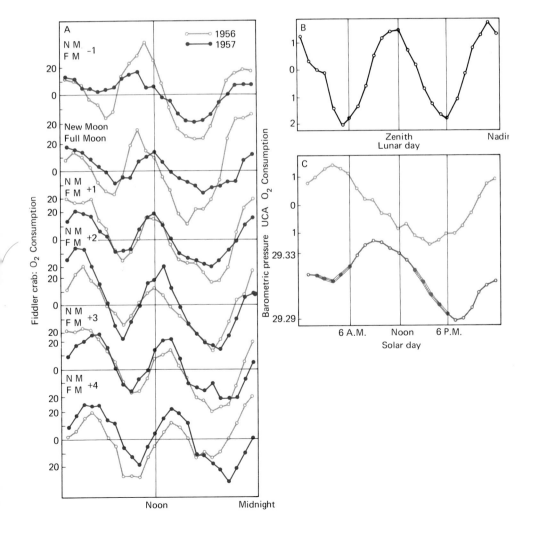

respirometers, correlated with high statistical significance with the same-day weather-associated distortions in the solar-day tides of the atmosphere in a manner (only 2-6 A.M. and 2-6 P.M.) comparable to that found for the potatoes.

For one year, 1954, something seemed radically to "go wrong" with our crab mean daily cycle despite the fact that we did nothing differently to the best of our knowledge.

Figure 2-22 illustrates our anomalous finding of 1954 together with the more typical one we had for the same month of the succeeding year. In July of 1954, we found a daily pattern that was not only essentially inverted but of a different form, despite the fact that the crabs had been collected from the same place and tested in exactly the same way in the laboratory as in other years. A year after obtaining these results we chanced to learn that Professor J. Simpson and his associates at the University of Chicago were troubled because they too had discovered significant anomalies in the mean daily intensity variations of primary cosmic radiation over the same time. We were generously loaned some of their data. Their findings for the two crab-months are plotted along with the metabolic ones in Figure 2-22. For both Julys, the daily cycles of radiation and metabolism essentially mirror-imaged one another even though both were changing strikingly between the two years. The lower box in the same figure shows the mean lunar-day metabolic pattern for August for a seaweed (*Fucus*) picked off the rocks near the laboratory at Woods Hole, Massachusetts. Note that this pattern parallels the mean *lunar-day* fluctuations in the primary cosmic-radiation intensity. The mean daily fluctuations of all the other organisms—rats, potatoes, salamanders, quahogs, oysters, etc.—that we studied over this period also either paralleled or mirror-imaged the fluctuations in primary cosmic radiation. Somehow these organisms were apprised of the fluctuations in intensity of the radiation. However, these were "nonsense" correlations since primary cosmic radiation does not reach the surface of the earth, and derivatives of it that do reach the surface do not fluctuate in any simple manner related to it. The physicists derived their information by monitoring the hourly fluctuations in neutrons and using these data to compute the primary-radiation intensity in the outer reaches of the atmosphere. Even though there obviously could be no direct cause-effect relationship between the radiation and organismic activity, it was evident beyond all reasonable doubt that these two quite different phenomena were not independent of one another. There had to be an explanation. It finally occurred to me that the fluctuations in intensity of primary cosmic rays entering the earth's atmos-

FIGURE 2-22

"Nonsense" correlations between fiddler crab and seaweed persistent respiratory fluctuations and primary cosmic radiation ones. All are mean variations for 31-day periods. During July of 1954 and the same month in 1955, both the daily metabolic rhythms of crabs and the intensity rhythms of primary cosmic radiation inverted, thus maintaining the virtual mirror image relationship between them. The lower box illustrates the direct relationship between the mean lunar-day rhythm of the primary cosmic radiation intensity and the respiration rate of the alga, Fucus *[Brown, F. A., Jr. (1969). Can. J. Bot.* **47**, *287-298].*

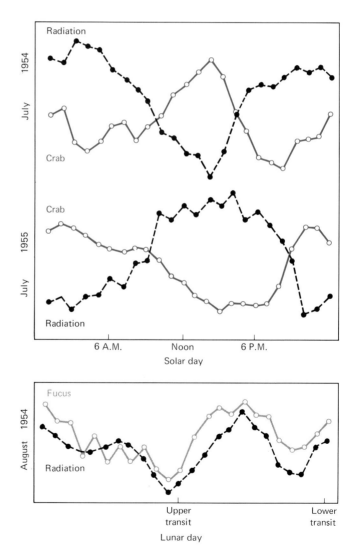

phere were dependent upon the strength of geomagnetism. The magnetic field steadily undergoes fluctuations in intensity. When the field is stronger, fewer primary cosmic rays come into the outer atmosphere; when it weakens, more get in. If organisms could be demonstrated to be sensitive to geomagnetism, the "nonsense" correlations between primary cosmic radiation and organismic metabolism or activity could be explained. An important question now arises: Do organisms have this sensitivity for geomagnetism?

Biologists for many years have pondered the question of the means by which birds, fish, and all sorts of other animals navigate over the face of the earth or find their way home when displaced from it. One postulated means that has been given substantial credence by the results from many experiments to test it is that a wide variety of organisms use the sun as a celestial reference for geographical direction. However, since the sun is not a stationary reference point, the birds must always "know" the time of day in order to "know" in what geographical direction the sun should be at that moment. For this information they are presumed to use their internal solar-day biological clocks (Figure 2-23). To complicate matters a little more, because of the geometry of the situation, the change in geographical direction of the sun with hour of the day is not uniform; at some times of day the sun moves faster over the compass, at other times slower. And to complicate matters even far more, the birds and other organisms appear equally able to use the moon and possibly even the star patterns to orient at night, suggesting that they must also possess lunar-day and sidereal-day clocks and use the correct one for each occasion.

We now began to wonder, if animals actually possessed the sensitivity for the geomagnetic field that the cosmic ray observations suggested, why had not the animals used the simple, direct response to these fields as references for geographical direction and saved themselves a great deal of complicated computing. Also, there are many reports that organisms can migrate, navigate, or home when the celestial references cannot be seen, even through fogs, storms, or at night. At any rate, in geographical orientation, biological clocks were related in an essential manner to the phenomenon. Orientations in time and space were interlocked phenomena.

The most promising point of attack upon the problem of the role of magnetism for biological clocks seemed to us to lie in this area of time-space relationships. To bring the "bird-problem" into the laboratory and to reduce it to practicable, workable dimensions, we devised the system illustrated in

FIGURE 2-23

Sun-compass orientation in a bird. To fly in a particular direction, birds are postulated to be able to use the sun as a celestial reference. Since the sun, however, is not a stationary reference, a bird desiring to fly, for example, due south must correct continuously for the sun's changing position to maintain a southerly compass heading. The bird is presumed to use its clock to implement this essential, steady alteration in the angle of its flight relative to the sun.

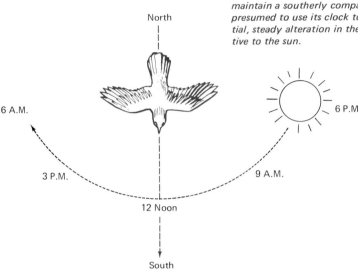

FIGURE 2-24

An experimental setup designed to measure the strength of turning response of the flatworm, Dugesia, to two light sources, one behind the worm and the other on its right. A flatworm is shown in the starting position. The turning from the right light for the worms is quantified by recording the point of worm crossing of the inner calibrated circle. Since the worms are negatively phototactic, the average values are always between 0 and -90°.

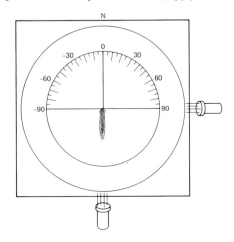

Figure 2-24. We replaced the bird with the tiny, ¼-inch flat-worm, *Dugesia*, and the celestial references with artificial lights. As seen in Figure 2-24, a polar grid was laid out under the bottom of a petri dish; the diameter of the grid was only 2 inches. Worms pointed toward the zero value on the grid were simultaneously illuminated from behind and from a second point-source to their right. Since these worms are negatively phototactic they tend to move away from the light and therefore their path usually turns into the upper left-hand quadrant (0 to -90°). As the worms cross the arc (which is the finish line) their angle is recorded, thus quantifying their turning response. The entire apparatus including its fixed lights may be pointed in any geographical direction so that one can learn whether the apparent response to lights reflects in any way the geographical directions of the lights or if for any fixed geographical arrangement of the lights the response to them varies with time.

When we directed the apparatus to the north and each morning obtained the average response from 45 worm-runs, we discovered that the angle varied with the day of the month. This is seen in Figure 2-25, summarizing the results from a four-year study. During the days centered around new moon, the left turning was significantly greater than over the days spanning full moon. Our assay system seems to be a good one, in that we are able to relate a biological clock—here a monthly one—to an orientation of an organism in space.

In the remainder of this chapter I shall complete the development of an argument for external timing, bringing in new key pieces for the case obtained through the use of the assay system illustrated in Figure 2-24.

Dr. Young Park and I embarked on a series of experiments. When the entire apparatus was directed north (Figure 2-26A) a mean turning of about 25.0° was obtained (Figure 2-26C). On the other hand, when the apparatus was directed toward the east (Figure 2-26B) the turning was about 29.0°. The standard errors, illustrated in the figure, point to the clear statistical significance of the differences, which can only be interpreted as follows: The animal was able to distinguish between the north and east headings. Since the total ambient environment of obvious factors for the organism was identical for the north and east orientations of the apparatus, the animals must be depending upon subtle information. We next ask: How do the worms distinguish the different directions? Searching for the answer to this question the following steps were taken. The apparatus was oriented toward the north but this time the field of a bar magnet was imposed on the experimental conditions.

FIGURE 2-25

*Synodic monthly variation in the degree of left-turning of initially north-directed flatworms tested in the late mornings. The mean value for each day of the month was calculated for each of four years and plotted in the figure. NM=new moon; FM=full moon. The worms turn more strongly away from the right light at new moon than at full moon [Brown, F. A., Jr. (1969). Can. J. Bot. **47**, 287-298].*

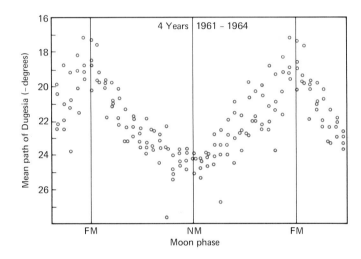

FIGURE 2-26

Experiments demonstrating that flatworms can use magnetic direction to orient themselves geographically. A and B show the apparatus oriented north and east, respectively, in the earth's magnetic field. The averages of the worm paths obtained in these two orientations are indicated in C by the terminal points of the dashed diagonal line. A' and B' show the apparatus remaining oriented to the north in both cases, but with the earth's magnetic field augmented about twentyfold by placing a bar magnet beneath the apparatus. The magnet was oriented to parallel the earth's field in A' but rotated 90° counterclockwise in B'. The average paths taken by the worms in these two conditions are indicated in C by the unbroken diagonal line. The size of the standard errors are portrayed by the vertical lines bisecting the points. Note that the worms act as if the whole apparatus had been rotated to the east in condition B', indicating that they can use magnetic direction to orient in space.

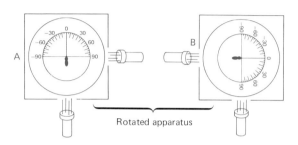

A   B

Rotated apparatus

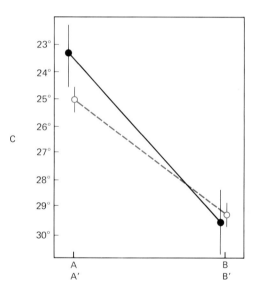

C

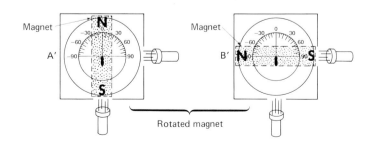

Magnet   Magnet

A'   B'

Rotated magnet

This magnetic field was applied with only a horizontal vector which was oriented in the same direction as the earth's field, but augmented to twentyfold (Figure 2-26A'). The mean path of the worms running in this condition was found to be 23.5° (Figure 2-26C). The larger standard error indicated that the behavior of the worms showed greater variability in this unnaturally strong field, but the mean path taken by the worms was not significantly different from the mean path obtained for north-directed worms in only the earth's field. When worm turning was measured with the apparatus still directed north but with the same magnet rotated a quarter-turn counterclockwise (Figure 2-26B') to give the organism the magnetic illusion that its field was now directed to east, the mean path was altered to 29.5°. Rotating the magnetic field counterclockwise by 90° relative to the apparatus appeared to be essentially the equivalent of rotating the apparatus 90° clockwise in the earth's field. The organisms were able to distinguish north from east by using the horizontal vector of a magnetic field. A living organism possesses the capacity to employ a subtle geophysical field component to distinguish geographical directions; the organism thus possesses a good magnetic compass.

Another kind of experiment demonstrated that this compass capacity is related to the biological clock mechanism. Each morning we headed the animals north (Figure 2-27A), establishing the form and phase of the worms' monthly turning rhythm; as usual the maximum turning to the left occurred at the time of new moon (Figure 2-27C). Immediately on completing the foregoing observations each day the apparatus was rotated 180° (Figure 2-27A') and again average turning was measured. As seen in Figure 2-27C, this treatment immediately

FIGURE 2-27

Altering by magnetic fields the phase of the monthly variation in turning of the flatworm, Dugesia. A and A' are the geographical orientations of the experimental apparatus used to determine the monthly variations in turning rate shown in the top two curves of C. The monthly patterns are 180° out of phase with one another. B and B' describe another experiment in which the geographical orientation of the apparatus remained the same, but with a magnet placed under B' to reverse the direction of the magnetic field and to give the worms the magnetic illusion that the apparatus had been turned to the south. As seen in the lower two curves of C, the monthly pattern is again shifted in phase by 180° by the reversed magnetic field. In C the solid circles and NM indicates the day of new moon, and open circles and FM the day of full moon.

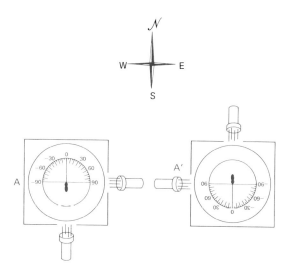

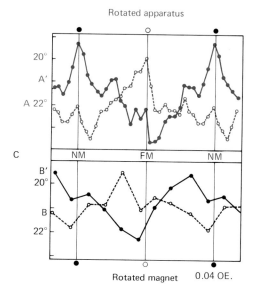

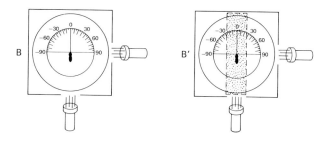

reset the organism's monthly rhythm by one-half cycle, i.e., 180°. These results suggest that the monthly turning rhythm is somehow dependent on a subtle geophysical factor that is related to geographical direction. To determine whether this phase change could be brought about by the same subtle factor, geomagnetism, by which the organism distinguished direction, we did the following. We directed the apparatus to the north (Figure 2-27B), establishing the form and phase of the monthly turning rhythm, and then at once repeated the observation after reversal of the magnetic field by an appropriately positioned bar magnet. In making these measurements some of the observations were made in the early afternoon though most were made in the morning. In this series the times of maximum and minimum turning appeared slightly displaced relative to the purely morning study when the apparatus had been similarly directed north. Of most importance to us here is that when a bar magnet reversed the magnetic field (Figure 2-27B') (with the ambient reversed field only a quarter the normal one) there was produced at once a 180° change in phase (Figure 2-27C) of the monthly rhythm. The 180° phase change in the monthly variation was effected by a 180° rotation of our lighting pattern relative to a magnetic vector, whether natural or artificial. A spatial vector component in the earth's subtle electromagnetic complex is clearly related to timing a biological clock-timed rhythmic behavior.

Finally, our experiments disclosed another capacity of a living system, an extraordinary one whose implications far transcend the phenomenon of "living clocks." We learned that the order in which we performed our experiments made a difference in the results we obtained. This is depicted in Figure 2-28 where the results shown in A and A', plus the top two

FIGURE 2-28

A persisting effect of previous treatment on the form of the monthly variation in orientation of flatworms. A and A' are the same as A and A' in Figure 2-27 and the upper two monthly patterns in C are the same as in Figure 2-27C except that one has been displaced by a semimonth. Note that the forms of the two patterns are closely similar to one another. B and B' indicate that the same orientations as A and A' were used again, but with the order reversed: the worms were first directed south, followed at once by redirection to north. As seen in the lower two curves of C the monthly pattern is very different from those of A and A', but similar to one another. The worms appear to "remember" the initial geographical relations of the lights and to respond to the lights in their new relations just as if the apparatus had not been reoriented.

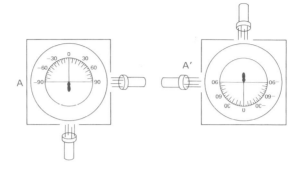

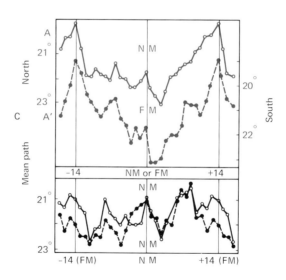

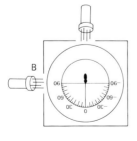

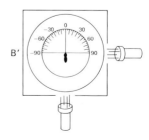

curves in C, are the same as those previously shown in Figure 2-27(A, A′, and C), except that here one of the monthly curves has been displaced by a semimonth to show that the two, though 180° shifted relative to one another, possess essentially the same form. New moon becomes the equivalent of the former full moon, and first quarter the equivalent of third quarter. Over the same period and on the same mornings exactly the same experiments were done except that the order was reversed: first south and then north (Figure 2-28B). Now the forms of the monthly cycles (Figure 2-28C) are quite different from those obtained under exactly the same conditions other than for the reverse sequential order. It is evident that the worms' monthly pattern of response to the asymmetrical illumination for one of the geographical orientations strongly tends to be retained even after that orientation has been altered. The "north" pattern persists even after the worms are redirected to south, and the "south" pattern of response persists even after the orientation has been redirected to north. The organism "remembers" the previous geographical direction of the light relations and continues to respond to them as if they still possessed the same geographical relations even after they no longer do so. The organism possesses the capacity to associate, and retain an association, between such an overt stimulus as light and subtle factors providing geographical directional information.

Whenever a rhythm approximating a natural geophysical frequency persists in constant conditions—thus indicating the operation of a so-called "biological clock"—there are always two possibilities for the timing of such a rhythm: (1) The organism may contain an independent, intrinsic timer acting like an hourglass in that it measures rather precise durations; or (2) the timing information may come from some subtle rhythmic geophysical parameter with the organism able to associate an event, or initiate a process, in relation to a particular, recognizable phase of this geophysical cycle. Return to the same phase in the recurring geophysical cycle could trigger the recurrence of the organismic cycle. One cannot establish the existence of an internal timing mechanism until one can exclude the possibility of an external one, and vice versa. It has been this ambiguity, incapable of resolution to date, that has permitted a controversy that has raged for many decades, a controversy that has generated far more heat than light.

The subtle geophysical parameters vary continuously in both space and time. The same geophysical parameters which vary with time of day, phase of moon, and time of year, also vary continuously in space at any given instant. Theoretically, the same parameters that can provide geographical compass information can also provide temporal information within the coordinates of the natural geophysical cycles. In Figure 2-29, I have represented space as a 360° compass circle, every point along which is fluctuating with time—illustrated by other circles such as the solar and lunar days. Diagrammatically we see here the space-time continuum. In Figures 2-26, 2-27, and

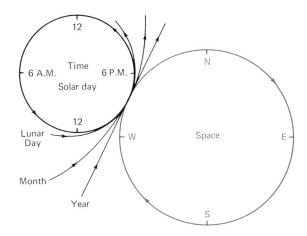

FIGURE 2-29

Subtle geophysical parameters vary continuously in space and time, forming a space-time continuum. At any instant in time there is a variation in space, here represented by the 360° geographical cycle. Each point in space is varying with time including solar-day, lunar-day, monthly, and annual periodic components. The demonstrations that (1) an organism can distinguish geographical directions by subtle geophysical parameters, (2) that involved parameters are related to the organisms' timing of their monthly rhythms, and (3) that organisms can "remember" directions of light stimuli in their relation to subtle parameters providing directional information suggest that an organism possesses just the appropriate capacities to "remember" light events in relation to the temporal cycles occurring in those same subtle parameters.

2-28 we have, in effect, transposed our ambiguous *temporal* problem to *space*. The ambiguity is gone.

Now we have discovered that animals are able to differentiate points along this 360° compass cycle by using subtle geophysical parameters alone, e.g., the horizontal vector of terrestrial magnetism as illustrated in Figure 2-26. Second, we have learned that when the magnetic field is rotated 180°, the flatworm's monthly turning rhythm is abruptly reset by half a cycle (Figure 2-27). In other words, a 180° rotation in a subtle spatial factor, 180°, resets a "biological clock." Subtle geophysical factors are thereby demonstrated to be intimately related to biological timing. Third, as was shown in Figure 2-28, an organism stimulated by light from a particular direction "remembers" (at least for many minutes) that a light had come from that particular geographical direction, doing so by associating the geographical pattern of illumination with subtle geophysical information that was providing geographical direction. In other words, the organism possesses the capacity to encode within itself information via obvious stimulus factors upon a 360° spatial cycle of variation.

The subtle geophysical characteristics that are unique to a particular compass direction, and with which the worm associated the light, are varying continuously in time with contained solar-day, lunar-day, monthly, and annual recurring patterns. Therefore, the organism has been shown to be able to associate light information with subtle factors that are varying with time. For example, viewing Figure 2-29, assume the organism has associated a light flash with the compass point (here shown as northwest by west where the time and space cycles become one). One can establish that the organism can distinguish this point on the spatial cycle because there is no time ambiguity, but in doing so we have simultaneously established that an organism can distinguish this point in time as well. In brief, since organisms have been shown to respond to specific geophysical directional information, and the latter information is varying cyclically with time, it is quite possible that these same subtle fields are also providing the organisms with their timing information. If the primary timing for organisms depends upon information steadily provided from the environment, then such remarkable properties of clock-timed rhythms as virtual independence of temperature and insensitivity to chemical disruptions become readily understandable.

In summary, the very fact that an organism possesses the three fantastic capacities described in Figures 2-26, 2-27 and 2-28—the very capacities that one might wish to develop if one were posed the problem of designing a clock system such as organisms are known to display—argues eloquently for this being the actual basis of the whole mysterious clock system, the solar-day, the lunar-day, the monthly, and the annual clocks. All could even be explained without invoking an independent internal timer for circadian or other periods. But let me emphasize this does not mean that these results prove that an organism *does not* have an independent internal biological clock system. Indeed, living things may actually have an internal timer as well, for redundancies are far from uncommon in living systems. These results, however, do point to a possible explanation of all the clocks of life which does not necessitate the postulation of a fully autonomous internal timer for these long periods, while at the same time enabling one to account for a number of reported properties of clock-timed rhythms that cannot be explained exclusively in terms of an internal timer. If an autonomous clock system truly exists in living creatures, one of the challenging areas for future researches will involve the nature and significances of physiological interactions between it and the environmentally dependent rhythmic system.

CELLULAR-BIOCHEMICAL
CLOCK HYPOTHESIS

J. WOODLAND HASTINGS

Professor of Biology
Department of Biological Science
Harvard University

The question of endogenous versus exogenous control of daily rhythms was debated long before it was explicitly recognized that these rhythms are implicated in the mechanism of biological time-keeping. Such mechanisms are now widely referred to as "biological clocks," the title of the first international symposium on the subject almost ten years ago in Cold Spring Harbor. I find it interesting to look back and quote from one of the opening paragraphs of the paper on biochemical aspects of rhythms which I read at that conference: "Today the biochemical approach is more concerned with mechanism. Although subtle refinements of the endogenous-exogenous questions are still discussed, all workers agree that an endogenous clocklike cellular mechanism exists, operating to bring about physiological and biochemical oscillations. The only consideration is whether the mechanism is relatively or absolutely autonomous. The recognized fact is that cell chemistry can and does oscillate, and the control involved is remarkable. This matter concerns us."

The status of the problem is in many ways very much the same today. A brief elaboration will perhaps help to provide some perspective.

Prior to 1954, when Dr. Colin Pittendrigh, then at Princeton University, elaborated the biological clock hypothesis with regard to persistent daily rhythms, the possibility of exogenous control of such rhythms had been frequently considered and debated, but not rigorously examined. This question was Bünning's primary concern in his early study on temperature effects on rhythmic sleep movements (Figures 1-1 and 1-2) in the bean plant, *Phaseolus*, in which he concluded that there was an endogenous mechanism which he likened to a metabolic "clock," since it "ran faster" at higher temperatures. Although he noted that the change in rate was relatively small compared to the effects of temperature on other metabolic processes, it was not until Pittendrigh's paper in 1954 that the broader biological significance of this feature was expanded upon.

Brown and Webb were similarly concerned with the endogenous-exogenous question in their early studies on the rhythms of color change in the crab *Uca*. Finding *complete* temperature independence in the period of the rhythm, they reasoned that it was difficult to assert unconditionally that the mechanism was wholly endogenous, since "one seeks in vain for descriptions of a [temperature independent] biological mechanism . . . [and] even with the existence of such a one it is difficult to credit it with sufficient precision . . . "

It turned out that *Uca* was apparently an exception; most if not all other rhythms examined have been found actually to *lack* precision in timing. The biological clock hypothesis, as it was further formulated and developed by Pittendrigh and his colleagues, capitalized on this by observing that selection pressure in the evolution of a true biological clock would be expected to operate only to match *approximately* the period of an endogenous cellular oscillator to that of the day—since the daily resetting signals of dawn and dusk would normally operate to correct for small inaccuracies. The "natural period" of a daily rhythm in an organism *not* exposed to the daily light-dark cycles is thus only approximately equal to one day (i.e., it is circadian; Figures 3-1 and 3-3). Moreover, it became clear that the exact length of a circadian period is itself not invariant—that it may be slightly different depending upon environmental conditions, or on the particular species studied; indeed, the length of a circadian period varies between individual organisms of the same species.[*]

From these and other considerations it became evident that the postulate of direct exogenous control, as it had been discussed and understood prior to the 1950's, could not be supported. The timing of a rhythm with a period which differs from 24 hours (and may be variable) can scarcely be attributed to a triggering cue with a fixed (e.g., 24-hour) period. The experiments shown in Figures 3-1 and 3-3 illustrate the circadian nature of the period. The details of this will be discussed later.

By 1960 a new hypothesis for exogenous control had been advanced by Brown. The hypothesis assumed an endogenous (and possibly autonomous) cellular clock mechanism possessing properties similar to those proposed by the endogenous school, but in addition the property of being sensitive to the hypothesized exogenous "geophysical cue," whose putative effect is that of providing time information of some sort.

Where does this leave us with regard to distinguishing the alternative theories? In the first place, the nature of the cellular biochemical clock mechanism should be of equal interest to proponents of both theories, since its properties are of crucial importance to both. But can we utilize our biochemical investigations to discriminate between the two theories? This is difficult to say, since the present hypothesis of exogenous timing provides no clear statement concerning the exact nature of the informational input to the system or the exact features

---

[*] There are small but significant differences in the period length of circadian rhythms in organisms kept at different temperatures, or at different levels of illumination. Drugs or chemical agents, notably (D_2O) (Figure 3-11) in the medium, may have similar effects. The small differences between individuals and different species are equally significant.

of the clock mechanism which are sensitive to or dependent upon the putative exogenous factors. One possible point of discrimination relates to the property of "temperature independence," which is viewed by proponents of the endogenous theory as a true feature of the biochemical mechanism; they hypothesized that the property is due to some sort of biochemical temperature-compensatory system, or is otherwise an inherent property of the biochemical oscillation. On the other hand, proponents of the exogenous theory have indicated that "temperature independence" is one of the features of the biological clock which is to be attributed to time information derived by the system from the external cue. A pursuit of this aspect of the problem may thus provide some distinctions of importance.

FIGURE 3-1

The phase relations of three rhythms in the dino-flagellate, Gonyaulax polyedra. *Dashed curve represents photosynthesis, as measured by the rate of ^{14}C uptake. The single acute peak represents the bioluminescent glow rhythm; the third curve, the stimulated bioluminescent rhythm (cells are induced to flash by mechanical agitation). All three rhythms will persist in constant conditions (although only the persistence of the latter is portrayed in this figure). Note that the period of this rhythm is not 24 hours as it is in the light-dark cycles, but becomes longer (24.2 hours) in constant dim illumination. Note also that the three rhythms are not in phase with one another in natural illumination cycles [Hastings, J.W. (1960).* Cold Spring Harbor Symp. Quant. Biol. **25,** *131-143].*

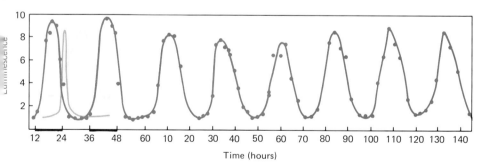

Let me mention one further point of importance with regard to environmental factors. The endogenous hypothesis does not hold that organisms are necessarily insensitive to any specific or unspecified physical factors; nor does it assert that the clock mechanism would necessarily be wholly independent of such factors, any more than it is independent of light or changes in light. It does assert that neither the special properties of the clock (e.g., temperature independence) nor its autonomous time-keeping ability are dependent on periodic changes in such factors. It is this comparison which thus distinguishes the two hypotheses most clearly at the present time.

The unicellular marine dinoflagellate *Gonyaulax polyedra* is a very favorable organism for the investigation of the biochemistry of circadian rhythms. Since many of the experiments I wish to discuss have been carried out with *Gonyaulax*, I should like to review briefly the rhythms involved. This organism possesses four different rhythmic systems that we have

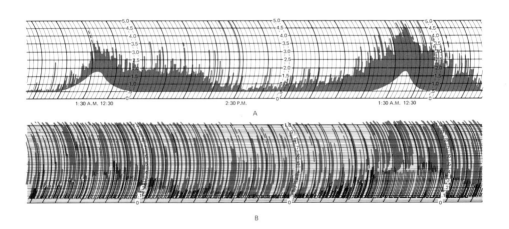

FIGURE 3-2

Two-day records of the luminescent glow rhythm, measured from one vial of cells kept in the dark (A), and from six vials kept in constant dim light (B) except for brief periods every hour when the vial was placed in a phototube chamber to measure its luminescence. The change in the baseline level represents the glow while the vertical lines are from spontaneous flashing of individual cells; temperature, 23°C [Hastings, J.W., (1960). Cold Spring Harbor Symp. Quant. Biol. **25**, *131-143].*

studied, and it seems likely that there may be more that could be followed if needed. This multiplicity of overt rhythms is in itself of considerable usefulness in studying mechanism; one can study how the several rhythms relate to one another and to the controlling mechanism.

The four different rhythms are illustrated in Figures 3-1 through 3-5. The flashing luminescence rhythm relates to the amount of light which can be obtained when the cells are stimulated. They characteristically emit light as individual fast (0.1-second) flashes upon mechanical stimulation; more and brighter flashing occurs during the night, and this can be measured quantitatively by integrating the light output during stimulation (Figure 3-1). Another luminescence rhythm is the glow rhythm (Figures 3-2 and 3-3). A relatively dim but still measurable glow may be noted toward the end of the normal

FIGURE 3-3

Summary of data taken from a vial of Gonyaulax *cells whose luminescence was measured for 20 days as described in Figure 3-2B. The time of day at which the maximum in luminescent glow occurred on each of the successive days is indicated by a triangular symbol. For the first five days the culture was kept on a daily light-dark cycle, and the peak occurred at the same time each day. Subsequently the cells were placed in constant dim light; the time at which the peak occurred was about 45 minutes later each day, so that the natural period was about 24 hours, 45 minutes; temperature, 24°C [from Hastings, J.W. (1960).* Cold Spring Harbor Symp. Quant. Biol. **25**, *131-145].*

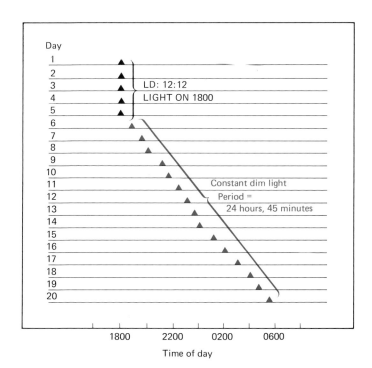

night. This lasts for only a few hours, and is usually accompanied by spontaneous flashing, which can be noted on the recordings as brighter spikes of light. The glow is recorded as the *intensity* of light. There is also a rhythm in the photosynthetic capacity of the cells, the maximum capacity occurring in the middle of the normal day (Figure 3-4). It may be assayed either manometrically or as the incorporation of radioactive carbon dioxide. Finally, there is a rhythm in cell division (Figure 3-5), which operates essentially like a gate, permitting those cells that are "ready" to divide to do so. In *Gonyaulax* the time of day when mitoses may occur falls toward the end of the normal night, close to the time when the "glow" occurs. Not *all* cells divide every day, but any which do divide during that day do so at that time. The result is a nonsynchronized but rhythmic population that exhibits a staircase-shaped growth curve.

FIGURE 3-4

Rhythms in photosynthesis and stimulatable bioluminescence in Gonyaulax *in alternating light-dark cycles. Dark periods (12 hours each) are indicated by black bars on the abscissa. The times at which cell divisions usually occur are indicated by the arrows. The curve labeled ^{14}C refers to measurements of photosynthetic capacity; the quantity of radioactive $^{14}CO_2$ incorporated when aliquots of cells were incubated with tracer in a saturating (960 footcandles) light intensity. Luminescence is in arbitrary units (right ordinates)* [Hastings, J.W. et al. (1961). J. Gen. Physiol. **45**, 69-76].

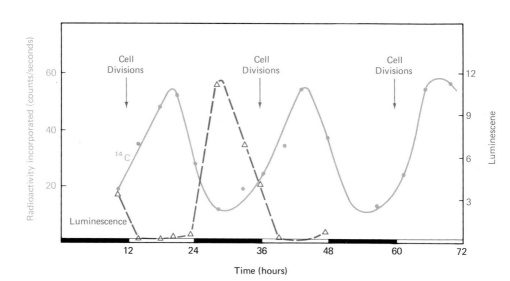

FIGURE 3-5

The circadian rhythm of cell division in Gonyaulax polyedra. The cells were grown photosynthetically with alternating light and dark periods of 12 hours each, the dark period being indicated on the abscissa by the bars. Cell divisions occurred predominantly at the end of the dark periods, so that there was a sharp increase in the cell number at about that time [Sweeney, B.M., and J.W. Hastings (1958). J. Protozool. 5, 217-224).

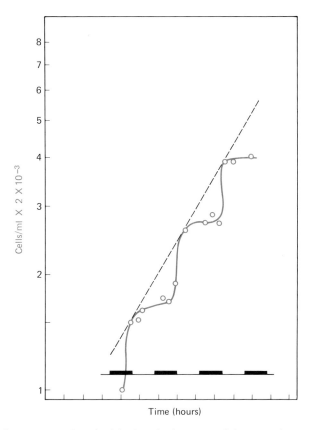

Time (hours)

Notions concerning the biochemical nature of the remarkable biological timing system have been long discussed, but experimentally neglected. Of the studies reported, the most extensive are those which involved the use of inhibitors and an analysis of their effects.

But a question which immediately arises when an effective inhibitor is discovered concerns its site of action. Most workers in the field of circadian rhythms have adopted a working hypothesis which distinguishes the mechanism that generates and controls rhythmicity from the target systems affected. An analogy with a clock and its hands has often been made; translating this into a hypothetical chemical mechanism (Figure 3-6), we can represent the clock mechanism by a reaction sequence in the form of a loop with appropriate feedback and circadian properties. One or more of the intermediates in this loop are appropriately coupled to specific biochemical systems which control overt rhythms (A, B, C, D, etc.). These might presumably correspond to the several overt rhythms in *Gonyaulax* described above. It is to be expected that the cycle will go faster at higher temperatures, by virtue of the fact that individual reaction rates are greater. How might feedback control be exerted in this hypothetical loop so that it would be substantially temperature independent? One of the simple models which we suggested some years ago, illustrated here in Figure 3-7, is that the step A to B, for example, was rate controlling in the loop and that an effector substance I (produced in a temperature-dependent reaction $H{\rightarrow}I$) exerted an inhibitory effect on this reaction, such that at higher temperatures more I

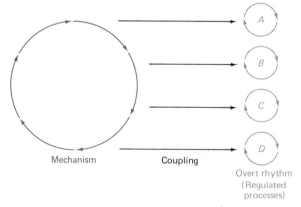

FIGURE 3-6

Schematic representation of a model for the hypothetical structure and steps involved in circadian rhythms. It is intended to signify an autonomous rhythmic mechanism which may be phase sensitive to appropriate light treatments, and to have special properties with regard to temperature dependency, and which drives the overt rhythms—the cellular phenomena A,B,C,D, etc.

was produced with the result that the actual rate of A to B was maintained rather constant. With such a model, one would predict that there might be instances where too much I was produced as the temperature was increased, thus actually slowing the rhythm down. In fact, such cases are known. As seen in Figure 3-8, which represents the luminescence rhythm in

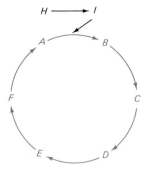

FIGURE 3-7

Diagrammatic representation of "mechanism," a chemical loop which, as it cycles, functions as the internal clock. A feedback action between H→I, and A→B, endows the loop with virtual temperature independence. Compensation is brought about as follows: both the reaction A→B and H→I are temperature dependent and the production of both B and I should increase as the temperature is raised. I, however, is a specific inhibitor of the reaction A→B. Therefore, as the temperature is raised, the increased production of I inhibits the rate of A→B from increasing, thereby stabilizing the entire loop from A to F and adequately buffering the clock mechanism against temperature change.

FIGURE 3-8

The effect of different constant temperatures on the luminescence rhythm of the dinoflagellate, Gonyaulax. *Prior to the start of this experiment the cells were kept at 22°C in alternating light-dark conditions. At the end of a dark period they were transferred to constant dim light (100 footcandles) and one of the test temperatures. Note that the periods lengthen as the temperature increases* [Hastings, J.W., and B.M. Sweeney (1957). Proc. Natl. Acad. Sci. U.S. **43**, 804-811].

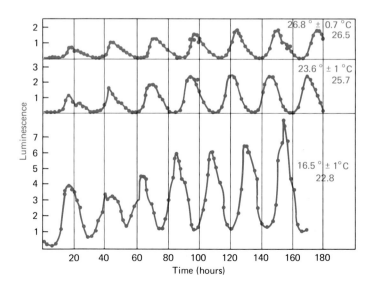

Gonyaulax, as the temperatures are increased up to 26.8°C, the clock apparently runs slower. The temperature coefficient of the response was 0.85. Note also that the amplitude is increased at lower temperatures—a very unusual phenomenon. If one plots the data as in Figure 3-9, it is seen that as the temperature increases so does the period, up to a point (27°C) where the effect is reversed.

FIGURE 3-9

The effect of different constant temperatures on the period length of the luminescence rhythm in the dinoflagellate, Gonyaulax. *The period is seen to* lengthen *with increasing temperatures up to 27°C, after which it decreases [Hastings, J.W., and B.M. Sweeney. (1957).* Proc. Natl. Acad. Sci U.S. *43, 804-811].*

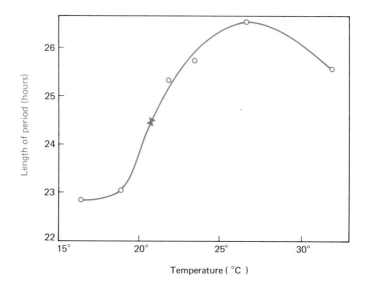

Length of period (hours) vs. Temperature (°C)

With reference to the scheme of Figure 3-6, it can be said that in the experimental sense the distinction between clock and hands is real or at least useful. It has been shown that there are rhythmic phenomena that can be completely blocked by using specific inhibitors without any substantial effect on the phase of the clock. For example, photosynthesis in *Gonyaulax* can be partially or completely blocked using the highly specific drug, dichlorophenyl dimethyl urea (DCMU); yet the glow rhythm of bioluminescence continues all the while. Furthermore, the rhythm of photosynthetic capacity will resume in constant conditions, *in phase*, following the removal of the inhibitor. A similar, though less readily interpretable reversible inhibition of the luminescent glow occurs in the presence of puromycin, a specific inhibitor of protein synthesis (Figure 3-10). This will be discussed more later on; the fact that the rhythm resumes in phase after puromycin removal means that the system which it inhibits to block the rhythm cannot be a member of the "mechanism" loop in Figure 3-6, but is more appropriately identified in that model with the biochemistry of an overt rhythm loop. This is appropriate for, as was true in the case of DCMU, there is evidence that other rhythms are not equally sensitive to puromycin.

From these and other similar observations it may be deduced from the model of Figure 3-6 that there is no feedback from the chemistry of the overt controlled process to the chemistry of the mechanism. Moreover, the notion that there exists a transducing mechanism between the two seems justified.

The search for true inhibitors of the hypothetical "mechanism loop" has not met with great success. Even the experimental technique needed to screen for such inhibitors is not altogether obvious. In principle, an inhibitor of the mechanism should not simply stop the rhythmicity; if it is reversible one should observe a phase shift in the rhythm equal to the number of hours of effective inhibition. As noted above in connection with DCMU and puromycin, an inhibitor of a target system, should, by contrast, have no such phase-shifting capability.

FIGURE 3-10

Diagrammatic representation of the effect of a single pulse of puromycin on the luminescent rhythm of Gonyaulax. *During the treatment, luminescence is inhibited; but when the puromycin is washed out the rhythm resumes, and the phase is unchanged, showing that the clock had been running all the time. As described later in the text, protein synthesis is only partially inhibited by the treatment.*

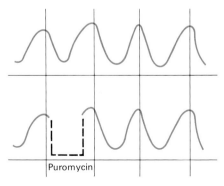

Time ⟶

In 1955, Dr. Fritz Bühnemann reported a careful study of the effects of certain inhibitors (cyanide, arsenate, fluoride, dinitrophenol, and several others) in an alga, *Oedogonium*, and found that both the phase and the period of its rhythm of sporulation were remarkably insensitive to these substances, even though the level of sporulation was substantially depressed. Bünning also reported extensive but largely negative experiments along these lines with *Phaseolus*. In more recent studies from his laboratory an interesting phase-shifting effect of alcohol on this rhythm was reported, but its mechanism of action has not been studied.

Another unusually interesting but still unexplained effect is that of D_2O, first reported by Drs. Victor Bruce and Colin Pittendrigh at Princeton. They found that the natural period of the phototactic rhythm of the flagellate *Euglena* was altered from about 23 hours to 27 hours when the water in the medium was largely replaced by D_2O. Drs. Bünning and Baltes, at the University of Tübingen, found that exposure of *Phaseolus* to D_2O caused a pronounced phase shift. In mammals, too, D_2O has apparently similar effects; Drs. Robert Suter and Kenneth Rawson at Swarthmore College reported that D_2O in the drinking water of the deer mouse increased the natural period of activity rhythm as a direct function of the deuterium concentration: at the maximum concentration (30%) the period was increased by 6% (Figure 3-11).

In 1960 our laboratory reported that a large number of inhibitors and other biologically active compounds were relatively ineffective in phase shifting in *Gonyaulax*. We used the "pulse" technique, in which the organisms were exposed to the compounds for periods of only about 8 hours. Although some effects were found, none was interpreted as having blocked a step in the mechanism. The most effective compound was arsenite; this has been confirmed in repeated experiments.

These and the previously mentioned results should be reviewed with the fact in mind that a relatively brief exposure to light *can* cause the phase of the rhythm to be shifted by many hours (Figure 3-12). Since the primary action of an absorbed photon must be photochemical, a light pulse may be legitimately viewed as a chemical pulse which *can* affect the hypothetical loop so as to cause a phase shift. The detailed features of this response will not be discussed here; suffice it to say, as can be deduced from the plot of Figure 3-13, that the number of hours by which a rhythm is shifted by a light pulse is different at different times during the cycle. Such response curves are incredibly similar in different organisms, ranging from the

FIGURE 3-11

Event-recorder tracings of persistent activity rhythms of the deer mouse, Peromyscus, *under treatment with D_2O. Successive days are shown starting at top. Percent D_2O and duration of treatment indicated on records. In all cases, D_2O increased the length of the period [modified from Suter, R.B., and K.S. Rawson (1968).* Science **160**, *1011-1014]*.

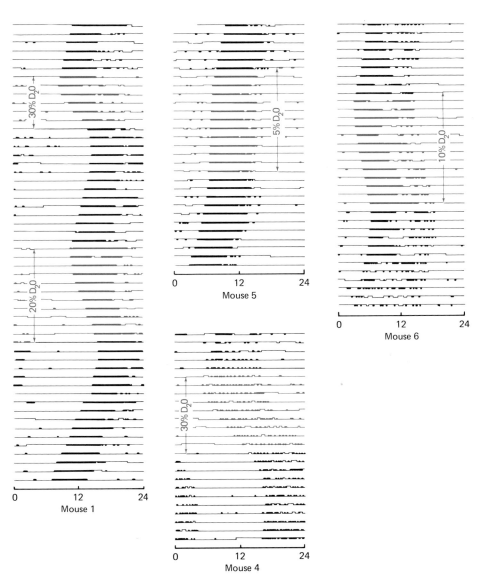

FIGURE 3-12

This experiment illustrates a phase shift in a bio-luminescent rhythm in Gonyaulax *following changes in light intensity. Cells which had been kept in a light-dark cycle were placed in constant dim light and under constant temperature conditions at the end of a 12-hour dark period. Two days later (zero time on the graph) measurements of the stimulated luminescence were begun and the circadian rhythm was apparent. Some cultures (A) were transferred to bright light (1400 foot-candles) for a period of 6 hours and then returned to the previous condition. Other cultures (B) were transferred to the dark for 6 hours and returned to constant dim light. The times at which the treatments were given are indicated by bars on the time axes. In both cases a marked phase shift in the rhythm is evident. The control (C) was left in dim light all the while; its natural period was 25.7 hours [Hastings, J.W., and B.M. Sweeney (1958).* Biol. Bull. **115,** *440-458].*

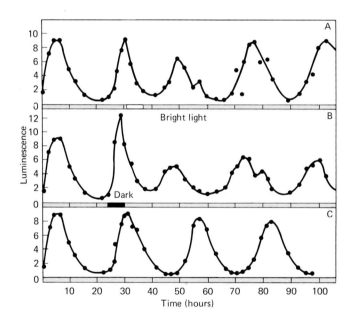

FIGURE 3-13

The relationship between the time in the cycle when a 2.5-hour light pulse (1400 footcandles) is administered and the resulting phase change. Phase shift is plotted on the ordinate as either a delay or an advance. Solid circles represent the stimulated bioluminescent rhythm and the open circles the luminescent glow rhythm. Note that pulses during the early part of the subjective night cause a phase delay and those during the latter part a phase

*advance [Hastings, J.W. (1964). In "Photophysi-
ology" (A. Giese, ed.) Vol. I, pp. 333-361.
Academic Press, New York].*

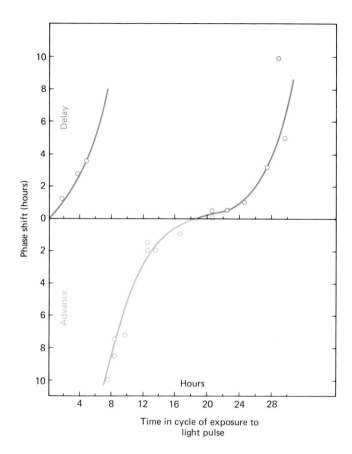

protozoa to mammals. The phenomenon is clearly significant
and it provides a valuable clue which should be capitalized
upon in future studies. Moreover, this feature must be suitably
incorporated in any clock model which may be proposed.

In 1962 we discovered that a number of antibiotics had
very profound effects on the rhythms in *Gonyaulax*, although—
unfortunately— none of these could be shown to cause a significant phase shift in any of the rhythms. The several inhibitors which were effective are those which act by blocking synthesis of certain macromolecules—ribonucleic acid (RNA) and
protein (Figure 3-14), but not deoxyribonucleic acid (DNA).
In fact, inhibitors of DNA synthesis (though effective, as
judged by the fact that they stop growth of the cells) were
found to be completely without effect upon the rhythms.

FIGURE 3-14

Effects of actinomycin D (concentrations as noted on the figure), chloramphenicol ($3 \times 10^{-4}M$), and puromycin (10^{-5} M) upon the persistent glow rhythm of bioluminescence. Top curve: control for experiments with actinomycin D. Bottom curve: control for experiments with puromycin and chloramphenicol. Cells were grown on a light-dark cycle and placed in constant dim light and constant temperature at the end of a light period, zero time on the graph. Inhibitors were added at the time indicated by the arrow. Luminescence in arbitrary units. Vertical grid lines are for guidance [Karakashian, M.W., and J.W. Hastings (1962). Proc. Natl. Acad. Sci. U.S. **48**, 2130-2137].

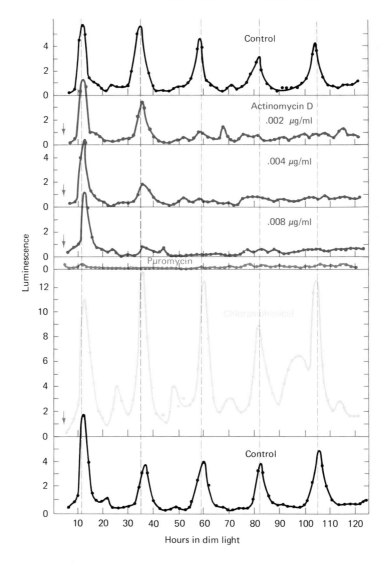

The postulate which emerged from these studies was one in which the hypothetical loop involves the synthesis of messenger RNA (mRNA). The most spectacular and specific effect was that of actinomycin D, a well-known specific inhibitor of RNA synthesis. At very low concentrations (0.1 μg/ml, or less) it completely blocked the luminescent glow rhythm, but only after one day. There appeared to be preformed material available to evoke and/or control one rhythmic cycle subsequent to inhibition by actinomycin. It seemed very attractive and appropriate to hypothesize that the circadian system should be compared to a special type of developmental system in which a daily differentiation was occurring. In fact, several authors were led to develop the hypothesis in greater detail, especially Dr. Brian Goodwin, at the University of Sussex, in his broad-ranging monograph on temporal control in biological systems more generally, and Drs. Ehret and Trucco at Argonne National Laboratory, in their highly picturesque "chronon" hypothesis (Figure 3-15). The latter authors developed a model in which they proposed that the DNA complement of a eukaryotic cells is comprised of functional subunits termed chronons, each of which is composed of a segment of DNA. The chronon is substantial in size, namely, polycistronic, and is functionally distinct because it corresponds to a daily scanning cycle: the chronon is regulated so as to get transcribed in a linear and temperature-independent fashion once a day. The message encoded at the termination of a chronon provides for reinitiation; whether all or only part of the DNA is organized into chronons was left vague, but a considerable number of chronons per eukaryotic cell were apparently envisioned.

The evident consequence of the chronon theory is that the synthesis of much if not all new cellular material will be programed on a circadian basis. This is an acceptable but surely insufficient hypothesis to account for the very spectacular differences which occur in circadian systems. The activities of specific circadian systems commonly vary by a factor of between 10 and 100 (Figure 3-1) over a time period when cell components increase by a factor of less than 2—usually far less than that. One would have to propose, presumably, that a complete (or substantial) turnover of *some* specific and critical components would occur on a daily basis, while others would be stable and receive only a daily increment. The chronon theory fails to address itself to this critical aspect of the problem.

We believe that "activation" followed by "inhibition" are key features of circadian control; while the effects may be mediated via RNA synthesis, it is difficult to account for the phenomenon on the basis of circadian timing of new synthesis of gross cellular components as such. In fact, it is not clear that this is a necessary part of any comprehensive theory.

The effects of inhibitors of protein synthesis merit a brief comment at this point. The results pictured in Figure 3-14 might appear to be somewhat contradictory, since puromycin

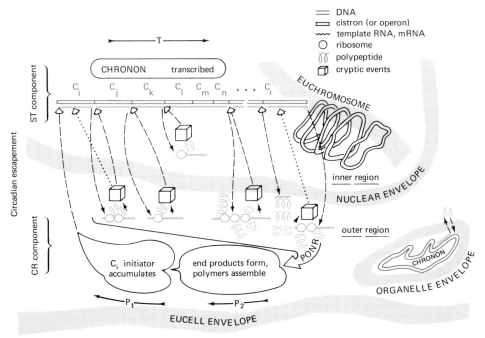

FIGURE 3-15

The chronon hypothesis of biological rhythmicity. The hypothetical chronon is a polycistronic complex of DNA. Synthesis of mRNA begins at one end and proceeds, cistron by cistron (essentially), along the chronon. Once formed, the mRNA from the first cistron diffuses out to the cytoplasm, and, on the ribosomes, directs protein synthesis. Some of these ribosomal products diffuse back into the nucleus and initiate the transcription of the next cistron on the chronon. This cycling between nucleus and cytoplasm continues until the end of the chronon is reached at which time an "initiator substance" accumulates which restarts the transcription cycle of the chronon. The entire process takes about 24 hours [Ehret, C., and E. Trucco (1967). J. Theoret. Biol. 15, 240-262].

completely blocks the rhythm, while in the presence of chloramphenicol (another inhibitor of protein synthesis), the rhythmic output is greatly amplified. These effects have never received a satisfactory explanation; it may be noted here that

in as yet unpublished studies by Laura McMurry at Harvard, it
has been found that protein synthesis (as measured by the
incorporation of radioactive amino acids) is not fully blocked
in *Gonyaulax* at the concentrations which produce the effects
pictured. Protein synthesis has also been implicated in clock
function in Dr. Jerry Feldman's study of the action of cycloheximide upon the phototactic rhythm in *Euglena.* The effect
was again not one of phase shifting, but involved a concentra-
tion-dependent change in the natural period of the rhythm
(Figure 3-16).

The various experimental findings with inhibitors indicate
that the clock mechanism, though resistant to chemical per-
turbation, is not entirely immune. Continued study can con-
fidently be expected to be fruitful. But in dealing with the
data we should probably not restrict ourselves to models so
simple as the single loop. The chronon hypothesis itself as-

FIGURE 3-16

The effect of cycloheximide on the period of the
persistent phototactic rhythm in the flagellate,
Euglena. *Graph shows direct potentiometer trac-*
ings; the tallest upright marks indicating strongest
phototactic response. C = controls; falling arrow
indicates time of cyclohexamide addition; concen-
trations (µg/ml) given at left of figure; days on
abscissa [modified from Feldman, J. (1967). Proc.
Natl. Acad. Sci. U.S. **57**, *1080-1087].*

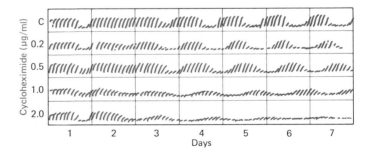

sumes single "mechanism" loops of the type represented in
Figure 3-6; as proposed the single chronon mechanism controls
a multiplicity of cellular processes by virtue of the transcription of different cistrons at different times of day. But since
the genome presumably comprises many chronons, a formal
model of the type shown in Figure 3-17 might be more appropriate, each an independent (or quasi-independent) "clock"
controlling its complement of "hands."

The question that is raised by these models has often been
discussed but never experimentally resolved, namely, is there

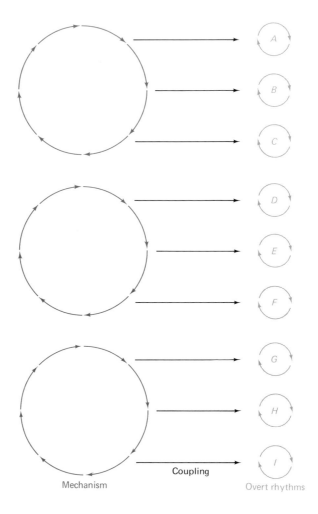

Mechanism   Coupling   Overt rhythms

FIGURE 3-17

An alternative to the model of Figure 3-6 in which more than one independent or quasi-independent mechanism is postulated.

a "master" clock mechanism or a multiplicity of clocks in a unicellular form? In some early experiments we were led to suggest the latter because of parallel temperature dependencies in the block and the luminescence intensity which the clock controls. A similar conclusion might be derived from the experiments with actinomycin D, which was found to be completely ineffective as an inhibitor of the rhythm of photosynthetic capacity while at the same time blocking completely (Figure 3-14) the luminescent glow rhythm. Unfortunately, the actinomycin D effect has not been found to be reversible, so it is not possible to draw firm conclusions from these particular experiments.

Attempts to dissociate the different rhythms in *Gonyaulax* in a definitive fashion have not been successful. Under conditions which cause changes in rhythms without obliterating any one, all rhythmic functions in a *Gonyaulax* cell culture have been found to be similarly affected. The choice between the models of Figures 3-6 and 3-17 thus cannot be made on the basis of present experiments.

But neither of these models is convincingly or at least irrefutably appropriate to the cell chemistry which may be imagined to be involved in circadian oscillations. We should probably not restrict ourselves to models so simple as the single loop. We know that there are a great number and variety of organelle and biochemical compartments in the cell; we know that biochemical pathways are replete with shunts, redundancies, and feedback. Biochemical oscillations may very well involve all of these in a way such that blocking any given pathway will have little or no effect on the observed cellular oscillation. The very interesting short-period oscillations both from intact cells and *in vitro* yeast extracts reported by Dr. Britton Chance and his colleagues at the University of Pennsylvania are instructive in this regard. Their analysis (Figure 3-18) indicates that even in a relatively simple system a complexity of shunt and feedback steps are involved. The oscillations of different components in the network possess different relative phase angles (Figure 3-19) such that the control of specific cell functions at specific times of the cycle could readily be envisioned. Models which involve principles of this type should be given more attention, both experimentally and theoretically.

Another matter worth some special discussion in connection with mechanism concerns circadian systems which are arrhythmic, i.e., those which lack any overt or detectable circadian components. Under such conditions, how can we envision the status of the "mechanism"—the operation of the hypothetical "loop" which generates circadian variations? Do these particular clock generating reactions simply cease? Or are they perhaps reactions derived from a specific cistron(s) which may have either a repressed or derepressed status? Or do the reactions continue but become uncoupled, using the model of Figure 3-6? Or do cellular clocks always continue to function, but the systems which are apparently arrhythmic being ones in

FIGURE 3-18

The sinusoidal oscillation of DPNH in a cell-free extract of the yeast enzyme system. The period is about 5 minutes. The amplitude of the curve represents about a 10% change. Also depicted is the pathway from glucose to ethyl alcohol with the consequent reduction and oxidation of DPN. [Chance, B., W. Pye, and I. Higgins. (1967). IEEE Spectrum 4, 79-86].

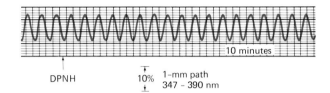

DPNH   10%   1-mm path   347 – 390 nm

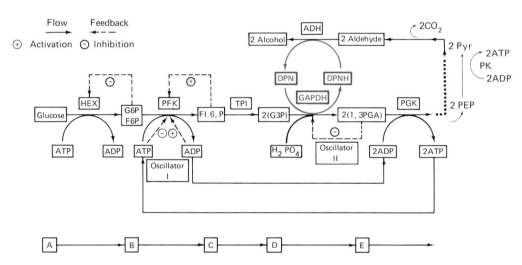

which the individual components have become asynchronous with one another? Or finally, if we visualize that circadian systems involve a chemical network with oscillations as discussed in the above paragraph, it might simply be that the arrhythmic system is one in which these *oscillations* do not occur even though the component reactions in the network continue.

The experimental facts, though meager, are interesting. Certain organisms, if raised from the egg (*Drosophila*) or the seed (*Phaseolus*) under absolutely constant laboratory conditions of light and temperature, fail to exhibit the typical circadian rhythm. Rhythmicity may be *initiated* in such systems by some sort of physical "trigger," such as a brief flash of light (even a millisecond may be adequate) or a temperature pulse (Figure 1-9). No physical *cycle* is required; the phase of the

FIGURE 3-19

A plot of the phase-angle relationships of glycolytic components and related compounds.

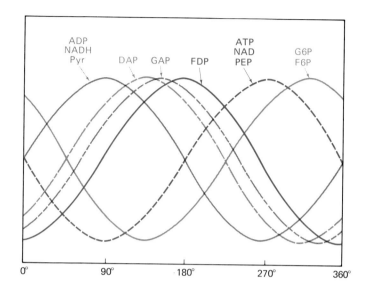

FIGURE 3-20

The initiation of an endogenous circadian rhythm of luminescence (stimulated flashing) in Gonyaulax *by means of a one-step change in illumination. The cultures had been grown in constant bright light (800 footcandles) and constant temperature for one year; no circadian components were detectable during that time; bioluminescence was invariant with time of day. At the time indicated on the graph as 0 hours, the cultures were transferred to dim light (90 footcandles), and a circadian rhythm ensued. Similar experiments in which the time at which the transfer to dim light was varied showed that the phase of the rhythm is determined by the time of the transfer. Natural period in this experiment, about 24.5 hours [Hastings, J. W., and B.M. Sweeney (1958). Biol. Bull. 115, 440-458].*

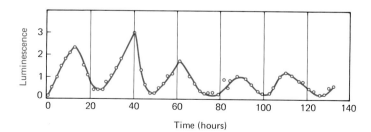

ensuing rhythm is determined by the time at which the triggering pulse was administered, while the period is an inherent property of the organismic system. This situation was hypothesized by Pittendrigh to constitute a case of asynchrony, in which the individual organisms (or cells) were rhythmic but out of phase with one another. The "initiation" pulse was viewed as a phase-shifting signal which, by virtue of the fact that cells in different phases respond differently (Figure 3-13), might serve to synchronize. However, it was evident from some of Pittendrigh's early experiments that a temperature pulse could *initiate* a rhythm, but could not phase shift a rhythm already "in motion." Bünning has cited several other distinctive differences between stimuli which initiate rhythms and those which phase shift, including differences in the action spectra for light stimuli. Thus the arrhythmic system seems to have a clearly distinctive physiological status.

Some circadian systems may be "inhibited" by constant light; the actual intensity of the light required to inhibit may vary with different organisms. Rhythms in such arrhythmic systems can invariably be "reinitiated" by simply transferring the organisms to a lower (noninhibitory) light level; the phase of the rhythm in this case also is determined by the time at which the light intensity is changed (Figure 3-20). Here too it seems clear that the arrhythmicity does not derive from asynchrony between cells in the population. Rhythms in individual isolated cells have been measured in *Gonyaulax* by Dr. Beatrice Sweeney and shown to behave as the population does, losing all overt rhythmicity in constant bright light.

A similar and possibly analogous loss or rhythmicity occurs in a number of systems at low temperatures. In *Gonyaulax* all overt rhythmicity is lost at temperatures below 13° C; the rhythm is reinitiated and phase determined upon increasing the temperature. Light intensity is maintained constant throughout.

Although rhythms in many systems may persist unabated for weeks or even months, there are numerous instances where rhythms have been noted to simply "damp out." Although studies of such systems have been neglected—for obvious reasons—there are good reasons to consider the phenomenon to be quite interesting, and to view the "degraded" state as arrhythmic and not asynchronous, in the sense discussed above. It would be very valuable to compare the biochemical status of an arrhythmic system with that of a circadian system.

These observations naturally lead one to pose some hard questions to any mechanism hypothesis. In the chronon theory, for example, what new rules for the timing of gene readout are applicable to such arrhythmic cases? How could one achieve nonrhythmic messenger production in a system where its synthesis is reputedly under the strict control of a sequential circadian machine? Equally difficult questions actually face any "mechanism-loop" theory.

It will also be evident that these particular observations are equally relevant to, and highly incompatible with, theories of exogenous control.

All of the above considerations are, of course, inextricably tied up with the question of the exact molecular basis for the overt rhythmic phenomena. Though much of the discussion about clock mechanism tacitly assumes that protein synthesis is involved, the extent to which this is so is not really known. It is not known how a rhythmic system is turned on and off, in the most elementary sense. What is it that oscillates? Is it enzyme quantity, enzyme localization, or enzyme activity? Does it relate to the synthesis of inhibitors and/or substrates, or perhaps to some kind of compartmentalization? The possibility that a subunit association-dissociation cycle occurs has been mentioned, or that tertiary structure changes may occur in the protein. It is probably fair to say that many workers have assumed that changes in enzyme activity or concentration are the key to the situation, but there are only a few instances where they have been shown to occur. Knowledge concerning these aspects of rhythmic systems could contribute very substantially to our understanding of the mechanism, and it is to be hoped that more studies will be directed toward a characterization of biochemical changes correlated with circadian rhythms.

Probably the most extensive work along these lines is with *Gonyaulax*, where changes in the extractable *in vitro* bioluminescent components have been found to occur along with the *in vivo* circadian rhythm. In *Gonyaulax* also, a change in the activity of the enzyme ribulose diphosphate carboxylase has been shown by Sweeney to occur in concert with, and quantitatively account for, the rhythm of photosynthetic capacity. She has concluded that these changes result from alterations in enzyme activity rather than *de novo* synthesis and destruction.

A different conclusion has been tentatively adopted in connection with the bioluminescent system, although the experiments are not yet completed to fully document this. The biochemical nature of the system is complex and not yet well understood; the brief account which follows should therefore be accepted as incomplete.

One might have thought that the biochemistry of a lumines-

cent enzyme could have been studied without encountering
excessive complications. Nothing could be farther from the
truth. It is now clear that one cannot conclude that an extracted
quantity accurately reflects the physiological condition in
the living cell. When the biochemistry of light emission in
Gonyaulax was first studied it was shown that extracts contain
soluble components which interact to produce light (the apparent enzyme and substrate being designated *Gonyaulax* luciferase and luciferin, respectively). It was quickly noted that the
luciferase activity was greater at night. Similar experiments
concerned with the substrate gave an unexpected result: the
maximum in extractable luciferin occurred during the day. This
apparent paradox was resolved following the observation that
the process of filtering the cells resulted in violent bioluminescence flashing during the night, but only a dull glow by day.
It was therefore hypothesized that differential substrate util-
ization was occurring. This idea was confirmed in experiments
in which flashing was inhibited prior to harvesting, whereupon
greater quantities of substrate could be isolated from cells at
night. The experimental result is reproduced in Figure 3-21.

The results finally obtained revealed a phase difference
between the peak of enzyme and substrate (Figure 3-22); this
has been confirmed and documented by McMurry. Because
enzyme *precedes* the substrate in time, the result does not sup-
port a hypothesis that circadian oscillations occur via a mechanism involving substrate induction of an enzyme. In that particular hypothesis enzyme induction was ultimately limited by
the fact that the enzyme itself destroys its inducer; assuming
further that the enzyme is itself unstable, the stage would be
set for the next cycle. Undamped oscillations can be shown to
be generated in such a system, but it is evidently not applicable to circadian rhythms.

The exact cause for the day-to-night difference in luciferase
activity has not yet been unequivocally established. The
present evidence indicates that the phenomenon is due to *de
novo* synthesis and destruction. McMurry has shown that the

FIGURE 3-21

*Graphic representation of the large differences in
the amounts and in the apparent rhythmic pattern
of substrate (luciferin) that can be extracted from
the cells, the differences depending upon the pretreatment of the cells by a brief exposure to a
higher temperature (34°C). The luciferin activity
which could be obtained from cells that received
no pretreatment was much lower and peaked in the
daytime rather than at night. The time of day*

(abcissa) represents the time of harvesting in relation to the environmental light and dark periods and does not relate the order in which the experiments were carried out [Bode et al. *(1963).* Science **141**, *913-915].*

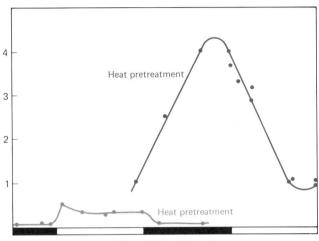

12-hour light—dark cycles

FIGURE 3-22

Diagrammatic representation of the amount of extractable enzyme activity and the amount of extractable substrate (luciferin) as related to time of day. The cultures were maintained on the light-dark cycle shown (12 hours of light, 12 hours dark). Cells were harvested and analyzed for luciferase and luciferin using the heat-treatment procedure described in Figure 3-21 for luciferin measurements. The bold arrow at the top of the graph indicates the time of day when the maximum in the glow of luminescence occurred in the cultures. The peak of substrate occurs somewhat later than the peak of enzyme.

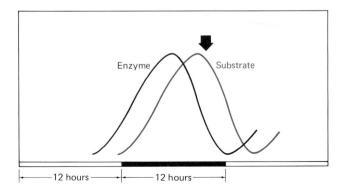

|←——— 12 hours ———→|←——— 12 hours ———→|

luciferase from cells grown in the presence of heavy isotopes will sediment more rapidly in a sucrose density gradient and can thereby be distinguished from the normal protein. Studies using this technique should permit her to distinguish *de novo* synthesis from alternative possibilities.

We assume that the control of bioluminescence *in vivo* involves more and/or different things than enzyme activity. Bioluminescence in *Gonyaulax* has now been shown to derive not from a simple enzyme-substrate system, but from a complex particle, possibly a cell organelle. This particle has been termed the scintillon, and it has been isolated in its active form and purified. The matter of concern may thus reduce—though the difficulties are not reduced—to the question of circadian control of the activity of this organelle. Such a situation further supports the notion that biological clocks involve physiological control mechanisms rather than simple enzyme synthesis and destruction. It thus underscores the view that biological control systems are in many instances concerned with a complicated target system whose control is subtle.

There are precious few other studies concerned with biochemical correlates in circadian systems to which I can even make reference. In our early enthusiasm we used to think that any and every aspect of cellular chemistry would display rhythmicity; quite the contrary seems to be the case, though even this relatively simple question is inadequately documented. Efforts to get at the heart of the mechanism have stimulated studies designed to demonstrate time-of-day specific RNA synthesis. Early studies in our laboratory using relatively insensitive techniques gave negative results. Ehret and Trucco have proposed and are undertaking DNA-RNA hybridization experiments to demonstrate that saturation of DNA by RNA produced at one time of day will leave sites that can be occupied by RNA produced at another time of day. It is to be hoped that some positive insights will result from those studies, but we would caution against overoptimism, since it is highly likely that even if the clock function does involve circadian production of RNA, there exists a large background of noncircadian RNA production as well. It is also to be hoped that more studies in more laboratories will be devoted to biochemical characterizations of circadian systems, especially those in single-celled organisms. A great improvement in our perspectives would result from improved biochemical knowledge.

It is true that the biochemical investigations which I have outlined fall short of dealing with the endogenous-exogenous question in an authoritative way. But the considerations which

I have outlined leave me and many of my colleagues with the conviction that biological systems can function autonomously as "clocks," based on a physicochemical cellular mechanism. The exogenous theory, while now accepting the existence of a cellular clock mechanism, rejects its fully autonomous nature. In the theory of exogenous control the physical nature of the hypothetical external influence has never been specified and any proof of the exogenous theory will require that the putative physical factors be specified and their properties described. In fact, in attempting to evaluate reports on the effects of exogenous factors on living organisms, it is frequently necessary to make extravagant concessions in order to pass over discrepancies which should have been experimentally resolved. The biological phenomena reputedly governed by exogenous factors display capricious behavior which is disconcertingly referred to as expected or unimportant, but remains unexplained. Attempts by other workers to repeat experiments or to reevaluate the published data using alternative statistical treatment have not resulted in a confirmation of the conclusion. In my opinion the hypothesis of exogenous control should be viewed with the greatest skepticism. But since this is a most useful and frequently fruitful way to view any hypothesis in science, we can only be optimistic about the future.

SELECTED READINGS

Aschoff, J. (1965). "Circadian Clocks." North-Holland Publ.,
Amsterdam.

Aschoff, J. (1963). Comparative physiology: diurnal rhythms.
Ann. Rev. Physiol. **25**, 581-600.

Aschoff, J. (1965). Circadian rhythms in man. *Science* **148**,
1427-1432.

"Biological Clocks" (1960). *Cold Spring Harbor Symp. Quant.
Biol.* **25**, 524.

Brown, F. A. (1954). Biological clocks and the Fiddler crab.
Sci. Am. April, pp. 34-37.

Brown, F. A. (1959). Living clocks. *Science* **130**, 1535-1544.

Brown, F. A. (1962). "Biological Clocks." Heath, Englewood,
New Jersey.

Brown, F. A. (1969). A hypothesis for timing of circadian
rhythms. *Can. J. Bot.* **47**, 287-298.

Bünning, E. (1967)."The Physiological Clock." Revised second
edition. Springer, Berlin.

Cumming, B.G., and E. Wagner (1968). Rhythmic processes in
plants. *Ann. Rev. Plant Physiol.* **19**, 381-416.

Ehret, C.F., and T. Trucco. (1967). Molecular models for the
circadian clock. I. The chronon concept. *J. Theoret. Biol.*
15, 240-262.

Goodwin, B. (1963). "Temporal Organization in Cells." Academic Press, New York.

Halberg, F., and A. Reinberg. (1967). Rhythmes circadiens et
rhythmes de basses fréquences en physiologie humaine.
J. Physiol. Paris **59**, 117-200.

Hamner, K. (1963). Endogenous rhythms in controlled environ-
ments. *In* "Environmental Control of Plant Growth"
(R. Evans, ed.), pp. 215-232. Academic Press, New York.

Harker, J. E. (1958). Diurnal rhythms in the animal kingdom.
Biol. Rev. **33**, 1-52.

Harker, J. E. (1961). Diurnal rhythms. *Ann. Rev. Entomol.* **6**,
131.

Harker, J. E. (1964). "The Physiology of Diurnal Rhythms."
Cambridge Univ. Press, London and New York.

Hastings, J. W. (1959). Unicellular clocks. *Ann. Rev. Microbiol.*
13, 297-312.

Hastings, J. W. (1964). The role of light in persistent daily
rhythms. *In* "Photophysiology" (A. Giese, ed.), pp. 333-
361. Academic Press, New York.

Mayersbach, H. von (1967). "The Cellular Aspects of Bio-
rhythms." Springer, New York.

Menaker, M. (1969). Biological Clocks. *BioScience,* **19,** 681-689.

Moore, S. (1967). "Biological Clocks and Patterns." Criteron, New York.

Palmer, J. D. (1970). Biological clocks. *In* "Encyclopedia of the Biological Sciences" (P. Gray, ed.). Second edition, pp. 107-108. Van Nostrand Reinhold, New York.

Rhythmic functions in the living system. (1962). *Ann. N. Y. Acad. Sci.* **98,** 753-1326.

Solberger, A. (1965). "Biological Rhythm Research." Elsevier.

Sweeney, B.M. (1963). Biological clocks in plants. *Ann. Rev. Plant Physiol.* **14,** 411-440.

Sweeney, B.M. (1969). "Rhythmic Phenomena in Plants." Academic Press, New York.

Symposium on Biological Chronometry. (1957). *Am. Naturalist* **91,** 129-195.

Webb, M. H., and F. A. Brown, Jr. (1959). Timing long-cycle physiological rhythms. *Physiol. Rev.* **39,** 127-161.